User's Guide to the Weighted-Multiple-Linear Regression Program (WREG version 1.0)

By Ken Eng, Yin-Yu Chen, and Julie E. Kiang

Techniques and Methods 4–A8

U.S. Department of the Interior
U.S. Geological Survey

U.S. Department of the Interior
KEN SALAZAR, Secretary

U.S. Geological Survey
Marcia K. McNutt, Director

U.S. Geological Survey, Reston, Virginia: 2009

For more information on the USGS—the Federal source for science about the Earth, its natural and living resources, natural hazards, and the environment, visit http://www.usgs.gov or call 1–888–ASK–USGS.

For an overview of USGS information products, including maps, imagery, and publications, visit http://www.usgs.gov/pubprod.

To order this and other USGS information products, visit http://store.usgs.gov.

On the cover: Examples of plots resulting from analysis of data with the weighted-multiple-linear regression (WREG) program (version 1.0).

Suggested citation:
Eng, Ken, Chen, Yin-Yu, and Kiang, J.E., 2009, User's guide to the weighted-multiple-linear-regression program (WREG version 1.0): U.S. Geological Survey Techniques and Methods, book 4, chap. A8, 21 p. (Also available at http://pubs.usgs.gov/tm/tm4a8.)

Contents

Figures

Tables

Conversion Factors

Multiply	By	To obtain
Length		
inch (in.)	2.54	centimeter (cm)
inch (in.)	25.4	millimeter (mm)
foot (ft)	0.3048	meter (m)
mile (mi)	1.609	kilometer (km)
mile, nautical (nmi)	1.852	kilometer (km)
yard (yd)	0.9144	meter (m)

User's Guide to the Weighted-Multiple-Linear Regression Program (WREG version 1.0)

By Ken Eng, Yin-Yu Chen,[1] and Julie E. Kiang

Introduction

Streamflow is not measured at every location in a stream network. Yet hydrologists, State and local agencies, and the general public still seek to know streamflow characteristics, such as mean annual flow or flood flows with different exceedance probabilities, at ungaged basins. The goals of this guide are to introduce and familiarize the user with the weighted-multiple-linear regression (WREG) program, and to also provide the theoretical background for program features. The program is intended to be used to develop a regional estimation equation for streamflow characteristics that can be applied at an ungaged basin, or to improve the corresponding estimate at continuous-record streamflow gages (henceforth referred to as simply gages) with short records. The regional estimation equation results from a multiple-linear regression that relates the observable basin characteristics, such as drainage area, to streamflow characteristics (for example, Thomas and Benson, 1970; Giese and Mason, 1993; Ries and Friesz, 2000; Eng and others, 2005; Eng and others, 2007a; Eng and others, 2007b; Kenney and others, 2007; Funkhouser and others, 2008).

The general multiple-linear regression for estimating a streamflow characteristic can be given by

$$y_i = \beta_0 + \beta_1 x_{i1} + \beta_2 x_{i2} + \ldots + \beta_k x_{ik} + \delta_i, \qquad (1)$$

where
y is the streamflow characteristic (dependent variable),
x_{ik} are basin characteristics (independent variables),
i (=1, 2, 3,..., n) is the index for gage i,
k is the number of basin characteristics,
$\beta_0, \beta_1, \beta_2,$ and β_k are the regression parameters, and
δ_i is the model error.

A critical issue in regional analyses is to understand the various sources of variability and error in the data. By understanding these sources, a user can select the appropriate approaches to estimate the regression parameters in equation 1, and the appropriate network of gages forming the region used to develop an estimate. Three approaches to estimate

regression parameters are provided in the WREG program: ordinary-least-squares (OLS), weighted-least-squares (WLS), and generalized-least-squares (GLS). All three approaches are based on the minimization of the sum of squares of differences between the gage values and the line or surface defined by the regression. The OLS approach is appropriate for many problems if the δ_i values are all independent of one another, and they have the same variance. Streamflow characteristics are estimated at gages using the available length of streamflow record. Because the length of record varies among gages, the precision of these estimates also varies, meaning that different δ_i values will have different variances.

A way to address the variation in the precision of estimated streamflow characteristics at each gage is to weight gages differently using WLS or GLS. A WLS approach reflects the precision of the estimated streamflow characteristic at that gage. An additional issue is that concurrent flows observed at different gages in a region exhibit cross correlations. If these correlations are not represented in a regional analysis, the regression parameters are less precise, and estimators of precision are inaccurate. A regional analysis that accounts for the precision of estimated streamflow characteristics, and the cross correlations among these characteristics, is known as GLS.

In addition to the variation and error in the data, the regression parameters in equation 1 are impacted by the choice of a network of gages forming a region. In a "conventional" regression, a region can be defined in several ways before a multiple-linear-regression study is initiated, such as by political boundaries or by physiographic boundaries. Within the context of "conventional" regressions, regions can also be defined during the regression study by using geographic information as an independent variable in the regression. Such regions can be defined using a variety of criteria, such as geographic grouping of similar residuals from an overall regression (Wandle, 1977), use of watershed boundaries (Neely, 1986), or physiographic characteristics. When performing a conventional analysis, a user of WREG must define the regions before using the program. WREG allows the user to perform conventional regressions using either OLS, WLS, or GLS.

An alternative to the conventional approach of pre-defining regions using political or physiographic boundaries is to define a region for each location of interest. This "region-of-influence" (RoI) regression approach defines a region and associated

---

[1]Former U.S. Geological Survey volunteer.

multiple-linear regression for every ungaged basin (for example, Acreman and Wiltshire, 1987; Burns, 1990; Tasker and others, 1996; Merz and Blöschl, 2005; Eng and others, 2005; Eng and others, 2007a; Eng and others, 2007b). A regression is formed on a subset of gages for which the values of independent variables are, by some measure, closest to those at the ungaged basin of interest. While the WREG program allows testing of RoI regressions, the application of RoI regression to ungaged basins must be accomplished using other programs, such as the National Streamflow Statistics (NSS) Program (Ries, 2006).

The first part of this report provides an overview of the multiple-linear regression techniques that are employed by WREG. It is followed by a step-by-step guide to the actual use of the program, including a description of the input files, the use of the graphical user interface, and an explanation of the output files.

Multiple-Linear Regression

In practice, the dependent variable, y_i, in equation 1 is an estimate, $\hat{y}_i$, often obtained from a limited sample size at each gage. The associated time-sampling error for the ith gage, η_i, is defined by

$$\eta_i = \hat{y}_i - y_i. \tag{2}$$

Substituting equation 2 into equation 1 gives

$$\hat{y}_i = \beta_0 + \beta_1 x_{i1} + \beta_2 x_{i2} + \ldots + \beta_k x_{ik} + \varepsilon_i, \tag{3}$$

where $\varepsilon_i = \delta_i + \eta_i$ (δ_i as given by equation 1).

The η_i values from gages close together will generally be correlated, because the finite sample of observed streamflows at one gage temporally overlaps the sample from another and temporal variations of streamflows are spatially correlated. Thus, the cross correlation between η_i and η_j for gage i and j will depend upon the cross correlation of concurrent flows at the two gages, and the number of concurrent years of record included in the dataset.

For a collection of gages with associated dependent and independent variables, equation 3 can be conveniently written in matrix notation as

$$\hat{Y} = X\beta + \varepsilon, \tag{4}$$

where

$$\hat{Y} = \begin{bmatrix} \hat{y}_1 \\ \hat{y}_2 \\ \vdots \\ \hat{y}_i \end{bmatrix} \quad X = \begin{bmatrix} 1 & x_{11} & x_{12} & \cdots & x_{1k} \\ 1 & x_{21} & x_{22} & \cdots & x_{2k} \\ \vdots & \vdots & \vdots & & \vdots \\ 1 & x_{i1} & x_{i2} & \cdots & x_{ik} \end{bmatrix} \quad \beta = \begin{bmatrix} \beta_0 \\ \beta_1 \\ \beta_2 \\ \vdots \\ \beta_k \end{bmatrix} \quad \varepsilon = \begin{bmatrix} \varepsilon_1 \\ \varepsilon_2 \\ \vdots \\ \varepsilon_i \end{bmatrix}, \tag{5}$$

and the total error, ε, is a random variable with a mean equal to zero and variance equal to σ_ε^2.

Independent and Dependent-Variable Transformations

The independent and dependent variables used in equation 5 can be transformed to obtain a linear relationship between the $\hat{Y}$ and X values. Common transformations include log (base 10), log (natural), and addition or subtraction of a constant. A user of WREG must choose appropriate transformations in the graphical user interface (GUI) before a multiple-linear regression is performed. A general transformation equation used by WREG is given as

$$V_{new} = f\left[\left(C1(V)^{C2} + C3\right)^{C4}\right], \tag{6}$$

where
V is the dependent or independent variable to be transformed,
V_{new} is the transformed independent or dependent variable,
f is either the log (base 10), log (natural), or exponential function, or a transformation can be omitted.
$C1$, $C2$, $C3$, and $C4$ are constants entered by the user.

Use of equation 6 for transforming variables is further discussed in the section of this report titled **Select Transformations**.

Estimation of Multiple-Linear-Regression Parameters

Following transformations of the dependent and independent variables, the transformed variables are used in WREG to estimate multiple-linear-regression parameters. When using any of the least squares regression approaches (OLS, WLS, or GLS), the regression parameters are estimated by

$$\hat{\beta} = (X^T \Lambda^{-1} X)^{-1} X^T \Lambda^{-1} \hat{Y}. \tag{7}$$

where X^T is the transpose of matrix X,
Λ^{-1} is the inverse of the weighting matrix Λ ($I = \Lambda^{-1}\Lambda$, where I is equal to the identity matrix).

The Λ matrix is constructed differently for OLS, WLS, and GLS, as described in the following sections. Once $\hat{\beta}$ is determined, it can be used to estimate the regression estimate of $\hat{y}$ at the ith gage, $\hat{y}_{iR}$, as

$$\hat{y}_{iR} = \hat{\beta}_o + \hat{\beta}_1 x_{i,1} + \hat{\beta}_2 x_{i,2} + \ldots + \hat{\beta}_k x_{i,k}. \tag{8}$$

However, a user should first check if the regression is adequate.

The estimators of Λ used in WREG for WLS and GLS approaches are applicable only to frequency-based streamflow characteristics. Alternative estimators of Λ to those presented in this manual can be explored using a user-defined option in

WREG (see section **UserWLS.txt**) for non-frequency-based streamflow characteristics, such as flow-duration exceedences.

Ordinary Least Squares (OLS)

For the OLS approach, $\boldsymbol{\beta}$ is estimated by (for example, Montgomery and others, 2001)

$$\hat{\boldsymbol{\beta}} = \left(\mathbf{X}^T \boldsymbol{\Lambda}_{OLS}^{-1} \mathbf{X}\right)^{-1} \mathbf{X}^T \boldsymbol{\Lambda}_{OLS}^{-1} \hat{\mathbf{Y}} = \left(\mathbf{X}^T \mathbf{X}\right)^{-1} \mathbf{X}^T \hat{\mathbf{Y}}, \quad (9)$$

where

$$\boldsymbol{\Lambda}_{OLS} = \mathbf{I} = \begin{bmatrix} 1 & 0 & \cdots & 0 \\ 0 & 1 & & \vdots \\ \vdots & & \ddots & 0 \\ 0 & \cdots & 0 & 1 \end{bmatrix}. \quad (10)$$

The OLS approach is suitable for estimating regression parameters when there is no variation in the precision of calculated dependent variables among gages, and the errors in equation 5 are independent of each other.

Weighted Least Squares (WLS)

For the WLS approach, $\boldsymbol{\beta}$ is estimated by (for example, Tasker, 1980)

$$\hat{\boldsymbol{\beta}} = \left(\mathbf{X}^T \boldsymbol{\Lambda}_{WLS}^{-1} \mathbf{X}\right)^{-1} \mathbf{X}^T \boldsymbol{\Lambda}_{WLS}^{-1} \hat{\mathbf{Y}}, \quad (11)$$

where $\boldsymbol{\Lambda}_{WLS}$ is the covariance matrix used to determine weights.

The components of the $\boldsymbol{\Lambda}_{WLS}$ matrix are a function of the type and source of the dependent variable. As with the OLS approach, the WLS approach is suitable when the errors in equation 5 are independent. However, for the WLS approach, weights in the weighting matrix are assigned so that gages that have more "reliable" estimates of streamflow characteristics have larger weights.

For streamflow characteristics calculated from a log-Pearson Type III frequency analysis (Bulletin 17B of the Interagency Advisory Committee on Water Data, 1982), Tasker (1980) provides a method for estimation of $\boldsymbol{\Lambda}_{WLS}$ that is used by WREG for this option:

$$\hat{\Lambda}_{WLS,ij} = \begin{cases} \sigma_{\delta}^2 + c_1\left(\dfrac{1}{m_i}\right) & (i = j) \\ 0 & (i \neq j) \end{cases}, \quad (12)$$

where

$$c_1 = \max\left[0, \ \bar{\sigma}^2\left(1 + \frac{\bar{K}^2}{2}\left(1 + 0.75\bar{G}^2\right) + \bar{K}\bar{G}\right)\right], \text{ and} \quad (13)$$

$$\sigma_{\delta}^2 = \max\left[0, \ \sigma_{OLS}^2 - c_1\left(\frac{1}{n}\sum_{p=1}^{n}\frac{1}{m_p}\right)\right], \quad (14)$$

where σ_{δ}^2 is the model-error variance,
m_i is the record length for the ith gage,
σ_{OLS}^2 is the observed mean-square error (MSE) of estimate using ordinary-least-squares approach,
$\bar{K}$ is the arithmetic average of the log-Pearson Type III deviates for all gages in the regression, and
$\bar{G}$ is the arithmetic average of the skew values at all gages (either at-gage skew, g, or weighted skew, G_w; explained below).

The log-Pearson Type III deviate values are a function of probability of exceedence and g (Interagency Advisory Committee on Water Data, 1982). $\bar{\sigma}$ is the arithmetic average of standard deviation of the annual-time series of the streamflow characteristic estimated by regression. This "sigma regression" is determined by OLS regression of the standard deviation of the annual-time series at each gage against basin characteristics at each gage (Tasker and Stedinger, 1989),

$$\sigma_i = \alpha_0 + \beta_{\sigma 1}x_{i1} + \beta_{\sigma 2}x_{i2} + \cdots + \beta_{\sigma k}x_{ik} + \varepsilon_{\sigma}, \quad (15)$$

where σ_i is the standard deviation of the annual-time series of the streamflow characteristic for the ith gage,
x_{ik} is the kth basin characteristic for the ith gage,
$\alpha_0, \beta_{\sigma 1}, \beta_{\sigma 2}$, and $\beta_{\sigma k}$ are parameters, and
ε_{σ} is the model error for the sigma regression.

The σ_i values are a required input into the WREG program as discussed in section **LP3s.txt**.

The weighted skew for the ith gage is given by (Bulletin 17B of the Interagency Advisory Committee on Water Data, 1982)

$$G_{w,i} = \omega_i g_i + \left(1 - \omega_i\right)G_{R,i}, \quad (16)$$

where $G_{R,i}$ is the regional skew estimate applicable to the ith gage, and

$$\omega_i = \left(\frac{MSE\left(G_R\right)}{MSE\left(g_i\right) + MSE\left(G_R\right)}\right), \quad (17)$$

where $MSE\left(g_i\right)$ is equal to the estimated mean square error of the skew value at the gage, and
$MSE\left(G_R\right)$ is the estimated mean square error of the regional skew values.

A variety of methods are available to determine G_R values (Bulletin 17B of the Interagency Advisory Committee on Water Data, 1982). Either g or G_w values are required input to WREG program as discussed in section **LP3G.txt**.

An alternative approach to calculating σ_{δ}^2 is presented by Stedinger and Tasker (1986). Their estimator is demonstrated to be more precise than equation 13, but their study did not include a mix of approaches to compute streamflow characteristics at partial-record-stream gages (Funkhouser and others, 2008). Use of equations 12 to 14 within the WREG program allows future versions to account for this mix of approaches for partial-record-stream gages.

Generalized Least Squares (GLS)

For streamflow characteristics calculated from a log-Pearson Type III frequency analysis, a GLS approach described by Stedinger and Tasker (1985) builds on the WLS approach by accounting for both correlated streamflows and time-sampling errors. This GLS approach estimates the β values by

$$\hat{\beta} = \left(X^T \Lambda_{GLS}^{-1} X \right)^{-1} X^T \Lambda_{GLS}^{-1} \hat{Y}, \tag{18}$$

where Λ_{GLS} is a matrix containing the estimates of the covariances of ε_i among gages.

The main diagonal elements of Λ_{GLS} thus include a part associated with the model error, δ_i, and all elements include the effect of the time-sampling error, η_i. In Tasker and Stedinger (1989), Λ_{GLS} is estimated by

$$\hat{\Lambda}_{GLS,ij} = \begin{cases} \sigma_{\delta i}^2 + \dfrac{\sigma_i^2}{m_i}\left[1 + K_i G_i + 0.5 K_i^2 \left(1 + 0.75 G_i^2\right)\right] & (i = j) \\[2em] \dfrac{\hat{\rho}_{ij}\sigma_i\sigma_j m_{ij}}{m_i m_j}\left[1 + 0.5 K_i G_i + 0.5 K_j G_j + 0.5 K_i K_j \left(\hat{\rho}_{ij} + 0.75 G_i G_j\right)\right] & (i \neq j) \end{cases} \tag{19}$$

where i and j are indices of locations of gages in the region of interest,

$\quad G_i$ and G_j are skew values equal to either g or G_w (equation 16) values for gages i and j,

$\quad m_i$ and m_j are record lengths for gages i and j,

$\quad m_{ij}$ is the concurrent record length for gages i and j, and

$\quad \rho_{ij}$ is an estimated value for the cross-correlation of the time series of flow values used to calculate the streamflow characteristic at gages i and j.

Values of the cross-correlation are estimated approximately by (Tasker and Stedinger, 1989)

$$\hat{\rho}_{ij} = \theta^{\left[\dfrac{d_{ij}}{\alpha d_{ij} + 1}\right]}, \tag{20}$$

where d_{ij} is the distance between gages i and j in miles, and

$\quad \theta$ and α are dimensionless parameters estimated from data as discussed in section **Model Selection**.

The $\sigma_{\delta i}^2$ values in equation 19 and the $\hat{\beta}$ values in equation 18 are jointly determined by iteratively searching for a nonnegative solution to (Stedinger and Tasker, 1985)

$$\left(\hat{Y} - X\hat{\beta}\right)^T \Lambda_{GLS}^{-1} \left(\hat{Y} - X\hat{\beta}\right) = n - (k+1). \tag{21}$$

Equation 19 does not account for error associated with estimating G. Depending on the actual magnitude of errors in estimation of skew, this additional error may unduly influence the estimation of β. Griffis and Stedinger (2007) proposed an approach to account for the uncertainty in the skew estimates in Λ_{GLS}, and this approach is used as an option by WREG. As implemented in WREG, this option assumes that weighted skews are provided by the user, and so this option should be activated only when weighted skews were used. The modified Λ_{GLS} matrix, $\Lambda_{GLS,skew}$, is given by

$$\hat{\Lambda}_{GLS,skew,ij} = \begin{cases} \begin{aligned} &\sigma_{\delta i}^2 + \frac{\sigma_i^2}{m_i}\left[1 + K_i G_i + 0.5 K_i^2\left(1 + 0.75 G_i^2\right) + \omega_i K_i \frac{\partial K_i}{\partial G_i}\left(3 G_i + 0.75 G_i^3\right)\right. \\ &\left. + \omega_i^2\left(\frac{\partial K_i}{\partial G_i}\right)^2\left(6 + 9 G_i^2 + 1.875 G_i^4\right)\right] + (1 - \omega_i)^2\,\sigma_i^2\,MSE(G_R)\left(\frac{\partial K_i}{\partial G_i}\right)^2 \end{aligned} & (i = j) \\[3em] \begin{aligned} &\frac{\hat{\rho}_{ij}\sigma_i\sigma_j m_{ij}}{m_i m_j}\left[1 + 0.5 K_i G_i + 0.5 K_j G_j + 0.5 K_i K_j\left(\hat{\rho}_{ij} + 0.75 G_i G_j\right)\right. \\ &\left. + 0.5\omega_i K_j G_i \frac{\partial K_i}{\partial G_i}\left(3\hat{\rho}_{ij} + 0.75 G_i G_j\right) + \omega_i \omega_j \sigma_i \sigma_j \frac{\partial K_i}{\partial G_i}\frac{\partial K_j}{\partial G_j}COV\left[g_i, g_j\right]\right] \end{aligned} & (i \neq j) \end{cases} \tag{22}$$

where $\dfrac{\partial K_i}{\partial G_i}$ and $\dfrac{\partial K_j}{\partial G_j}$ are the partial derivatives for gages i and j calculated from the Kite (1975; 1976) approximation for K given as

$$\frac{\partial K}{\partial G} = \frac{\left(z_p^2 - 1\right)}{6} + \frac{\left(z_p^3 - 6z_p\right)G}{54} - \frac{\left(z_p^2 - 1\right)G^2}{72} + \frac{z_p G^3}{324} + \frac{5G^4}{23{,}328}, \quad (23)$$

where z_p is the standard normal deviate corresponding to probability p.

The $COV[g_i, g_j]$ in equation 22 is the covariance between the skew values at gages i and j, and is given by

$$COV\left[g_i, g_j\right] = \rho_{g_i g_j} \sqrt{Var\left(g_i\right) Var\left(g_j\right)}, \quad (24)$$

where $\rho_{g_i g_j}$ is estimated by (Martins and Stedinger, 2002)

$$\hat{\rho}_{g_i g_j} = \frac{m_{ij}}{\sqrt{\left(m_{ij} + m_i\right)\left(m_{ij} + m_j\right)}} \text{Sign}\left(\hat{\rho}_{ij}\right)\left|\hat{\rho}_{ij}\right|^3, \text{ and} \quad (25)$$

$\text{Sign}\left(\hat{\rho}_{ij}\right)$ is equal to one if $\hat{\rho}_{ij}$ is positive and to minus one if $\hat{\rho}_{ij}$ is negative.

The $Var(g_i)$ and $Var(g_j)$ in equation 24 are approximated by (Griffis and Stedinger, 2009)

$$Var\left(g_i\right) = \left[\frac{6}{m_i} + a\right]\left[1 + \left(\frac{9}{6} + b\right)G_{R,i}^2 + \left(\frac{15}{48} + c\right)G_{R,i}^4\right], \quad (26)$$

where

$$a = -\frac{17.75}{m_i^2} + \frac{50.06}{m_i^3}, \quad (27)$$

$$b = \frac{3.92}{m_i^{0.3}} - \frac{31.1}{m_i^{0.6}} + \frac{34.86}{m_i^{0.9}}, \text{ and} \quad (28)$$

$$c = -\frac{7.31}{m_i^{0.59}} + \frac{45.9}{m_i^{1.18}} - \frac{86.5}{m_i^{1.77}}. \quad (29)$$

Equation 19 is a simplified version of equation 22 that assumes the skew is without error. Equation 19 is provided in the WREG program to reproduce previous studies that do not use equation 22.

Performance Metrics

The WREG program reports multiple performance metrics for multiple-linear regressions, depending upon the options (OLS, WLS, or GLS) selected. Specific metrics are reported either in the GUI or the output files of WREG.

Model- and Time-Sampling Errors

For conventional OLS and RoI regressions, the residual errors, e_i, are computed as

$$e_i = \hat{y}_i - \hat{y}_{ir}, \quad (30)$$

where $\hat{y}_{ir}$ is the estimated $\hat{y}_i$ provided by the regression (see equation 8).

The residual mean square-error (MSE) is computed as

$$MSE = \frac{1}{\left(n - k - 1\right)}\sum_{i=1}^{n}\left(e_i\right)^2, \quad (31)$$

where e_i is calculated from equation 30.

The MSE metric does not distinguish the proportion of total error, ε_i that is composed of model error, δ_i and time-sampling error, η_i. WLS and GLS regression provide estimates of the model error variance, σ_δ^2, which is the same as the MSE only if the time sampling error variance, σ_η^2, is equal to zero.

For conventional regressions using WLS and GLS, WREG reports the average variance of prediction, AVP, as the performance metric (Tasker and Stedinger, 1986) and is given by

$$AVP = \sigma_\delta^2 + \frac{1}{n}\sum_{p=1}^{n} x_p \left(X^T \Lambda^{-1} X\right)^{-1} x_p^T, \quad (32)$$

where x_p is a vector containing the values of the independent variables of the pth gage augmented by a value of one.

When $\hat{Y}$ corresponds to the logarithm of the variable of interest, equation 32 can be reported as a percentage of the predicted value. When expressed in this way, the metric is known as the average standard error of prediction, S_p (Aitchison and Brown, 1957, modified for use of common logarithms), and is given by

$$S_p = 100\left\{e^{\left[(\ln 10)^2 AVP\right]} - 1\right\}^{\frac{1}{2}}. \quad (33)$$

The standard model error as a percentage of the observed value can be calculated by substituting σ_δ^2 for AVP in equation 33. WREG program reports both S_p and the standard model error for WLS and GLS regressions.

For RoI regression, a regression is developed for each ungaged basin of interest. An overall performance metric reported by WREG program for RoI regressions using OLS, WLS, and GLS is a root mean square error, $RMSE(\%)$. This metric is similar to, but not the same as the prediction error sum of squares, $PRESS$, performance metric (for example, Montgomery and others, 2001). Every gage is treated in turn as an ungaged basin and a regression is developed for that site, and equations 30 and 31 are used to calculate a mean-square error value, MSE_{RoI}, that is used in place of AVP in equation 33 to give a root mean square error of prediction expressed as a percentage of the observed value, $RMSE(\%)$, given by (Eng and others, 2005; Eng and others, 2007a)

$$RMSE(\%) = 100\left\{e^{\left[(\ln 10)^2 MSE_{RoI}\right]} - 1\right\}^{\frac{1}{2}}. \quad (34)$$

Coefficient of Determination, R^2, R^2_{adj}, and R^2_{pseudo}

A metric reported by WREG for determining the proportion of the variation in the dependent variable explained by the independent variables in OLS regressions, is the coefficient of determination, R^2, (Montgomery and others, 2001) given as

$$R^2 = 1 - \frac{SS_r}{SS_T}, \tag{35}$$

where

$$SS_r = \frac{1}{n}\sum_{i=1}^{n} e_i^2, \text{ and} \tag{36}$$

$$SS_T = \sum_{i=1}^{n}(\hat{y}_i - \bar{y})^2, \tag{37}$$

where $\bar{y}$ is the arithmetic mean of all $\hat{y}$ values,

SS_T is the total sum of squares that is equal to the sum of the amount of variability in the observations, and

SS_r is the residual sum of squares.

SS_T and SS_r values are provided as WREG output files and can be used to calculate an adjusted coefficient of determination, R_{adj}^2, given as

$$R_{adj}^2 = 1 - \frac{SS_r/(n-k-1)}{SS_T/(n-1)}. \tag{38}$$

The adjusted R_{adj}^2 adjusts for the number of independent variables used in the regression.

For WLS and GLS regressions, a more appropriate performance metric than R^2 or R_{adj}^2 is the R_{pseudo}^2 described by Griffis and Stedinger (2007). Unlike the R^2 metric in equation 35 and R_{adj}^2 in equation 38, R_{pseudo}^2 is based on the variability in the dependent variable explained by the regression, after removing the effect of the time-sampling error. The R_{pseudo}^2 is given as

$$R_{pseudo}^2 = 1 - \frac{\sigma_\delta^2(k)}{\sigma_\delta^2(0)}, \tag{39}$$

where $\sigma_\delta^2(k)$ is the model error variance from a WLS or GLS regression with k independent variables, and

$\sigma_\delta^2(0)$ is the model error variance from a WLS or GLS regression with no independent variables.

For RoI regressions using OLS, WLS, or GLS, no coefficient of determination values are reported in WREG.

Leverage and Influence Statistics

The leverage metric is used to measure how far away the value of one gage's independent variables are from the centroid of values of the same variables at all other gages. This metric is reported for all gages in OLS, WLS, and GLS

regressions and RoI regressions. For non-RoI regressions, leverage, h, for the ith gage is given as

$$h_{ii} = \left[\mathbf{X}\left(\mathbf{X}^T\Lambda^{-1}\mathbf{X}\right)^{-1}\mathbf{X}^T\Lambda^{-1} \right]_{ii}, \tag{40}$$

where Λ is equal to either Λ_{OLS}, Λ_{WLS}, Λ_{GLS}, or $\Lambda_{GLS,skew}$.

The leverage metric for RoI regressions using either OLS, WLS, or GLS is given by (Eng and others, 2007b)

$$\mathbf{h}_0^T = \left[\mathbf{x}_0\left(\mathbf{X}^T\Lambda^{-1}\mathbf{X}\right)^{-1}\mathbf{X}^T\Lambda^{-1} \right]. \tag{41}$$

where $\mathbf{x}_0$ is a vector of independent variables at a particular ungaged basin.

Unlike equation 40, equation 41 measures the leverage that each gage record in the RoI has on the ungaged basin 0. Leverage metrics associated with gages are considered large if these metrics exceed the criteria given by

$$h_{limit} = \frac{C_h}{n}\sum_{i=1}^{n} h_{o,i}, \tag{42}$$

where C_h is a constant

For conventional regression, C_h is equal to 2 in equation 42 and reflects the observation that values twice the average can be considered as unusually large. For RoI regression, C_h is equal to 4 (Eng and others, 2007b). A larger C_h value for RoI regression is recommended because $\mathbf{h}_0^T$ values typically exhibit greater variability, and even negative values, whereas h_{ii} values are always positive. The h_{limit} is reported for regressions using OLS, WLS, and GLS in the output files of the WREG program, and is also shown visually on a plot of leverage values for each gage (see sections **Output Files** and **Leverage Values Versus Observations**).

The leverage metrics in equations 40 and 41 identify gages whose independent variables are unusual. Such unusual gages may or may not have any significant impact on the estimated regression parameters in equations 9, 11, and 18. An influence metric, such as Cook's D (Cook, 1977), indicates whether a gage had a large influence on the estimated regression parameter values. A generalized Cook's D value for the ith gage is

$$\text{Cook's } D_i = \frac{e_i^2 L_{ii}}{v\left(\Lambda_{ii} - L_{ii}\right)^2}, \tag{43}$$

where v is the dimension of β,

Λ_{ii} is the ith main diagonal of the Λ covariance matrix, and

L_{ii} is the ith main diagonal of $\mathbf{X}(\mathbf{X}^T\Lambda^{-1}\mathbf{X})^{-1}\mathbf{X}^T$ (Tasker and Stedinger, 1989).

For regressions, a gage that has caused large influence is identified if Cook's D exceeds the limit given by

$$D_{\text{limit}} = \left(\frac{4}{n}\right). \qquad (44)$$

The Cook's D limit calculated using equation 44 is reported in the WREG output files (see section **Output Files**), and visually on a plot of influence values (see section **Influence Values Versus Observation**).

Significance of Regression Parameters

Regression parameters are tested for significance by WREG for regressions not using RoI regions. The null hypothesis is generally that the regression parameter is equal to zero, and the alternative hypothesis is that this parameter is not equal to zero. If the null hypothesis is not rejected at some predefined level of significance, such as 5%, then the associated independent variable is removed from the regression. For regressions, the test statistic that is used to evaluate the null hypothesis is the *T value* statistic given as

$$T\ value = \frac{\beta_k}{\left(Var\beta_k\right)^{1/2}}, \qquad (45)$$

where $(Var\,\beta_k)$ is the covariance value of β_k, and is given as

$$Var\beta_k = \left(\mathbf{X}^T\,\mathbf{\Lambda}^{-1}\,\mathbf{X}\right)^{-1}_{kk}. \qquad (46)$$

The *T value* statistic in equation 45 is assumed to follow a Student's t distribution, so probabilities or "p-values" can be calculated. If the critical level of significance is set to 5% (0.05), p-values associated with the calculated *T value* statistic from equation 45 that exceed 0.05 result in acceptance of the null hypothesis. In this case, the regression parameter is not considered significant. In WREG, coefficients that are not significant at the 5% significance level are flagged (see section **Regression Summary**).

Definition of Regions

The region formed by a collection of gages can influence the regression model and the significance of the results. For conventional regressions, the WREG program uses all gages that are input by the user to develop regressions using either OLS, WLS, or GLS. If a subset of gages from a larger data set is desired by the user, input files to WREG should be modified to reflect the smaller subset.

For RoI regressions, a select subset from all available gages is formed for every ungaged basin where an estimate is desired. To use the RoI regression feature in WREG, a user would input all possible gages of interest, even though some gages might not be used in each RoI regression. Three approaches for defining hydrologic similarity among basins are available with WREG: independent or predictor-variable space RoI (PRoI) (for example, Burns, 1990), geographic space RoI (GRoI), and a combination of predictor-variable and geographic spaces called hybrid RoI (HRoI) developed by Eng and others (2007a).

For the GRoI option in WREG, a region of influence is formed using the n gages that are geographically the closest to the ungaged basin. In general, n is specified and is the same for every regression. The PRoI option in WREG is similar, but forms a region using the n closest gages in independent-variable space rather than geographic space. Thus, the region is comprised of gages whose independent variables have values that are the most similar to the ungaged site. Distance in independent-variable space from the ungaged basin to the ith gage, R_i, is defined in a Euclidean sense as (for example, Burns, 1990)

$$R_i = \left[\left(\frac{x_1 - x_{1i}}{s_{x_1}}\right)^2 + \left(\frac{x_2 - x_{2i}}{s_{x_2}}\right)^2 + \cdots + \left(\frac{x_k - x_{ki}}{s_{x_k}}\right)^2\right]^{1/2}, \qquad (47)$$

where s_{x_1}, s_{x_2}, and s_{x_k} are the sample standard deviations of x_1, x_2, and x_k, respectively (computed from data from the entire study region).

The HRoI approach uses a region of influence of the n closest gages in independent-variable space chosen from a subset of all gages having a geographic distance less than D from the ungaged basin. However, if fewer than n gages are available within the distance D of a given ungaged basin, then the limit D is ignored, HRoI reverts to GRoI and the n geographically closest gages are used. Thus, in the limit as D approaches zero, HRoI reduces to GRoI, and in the limit as D becomes arbitrarily large, HRoI reduces to PRoI (Eng and others, 2007a).

The user must determine the optimal D and n values. This determination can be accomplished by splitting the dataset into three equally sized subsets. Two of the three subsets are combined and used in an optimization step to calculate *RMSE* values for various values of n and D for HRoI regionalization. The same subsets are used to calculate *RMSE* values for various values of n for both PRoI and GRoI. The lowest resulting values of *RMSE* and the corresponding values of n and D for HRoI and of just n for PRoI and GRoI are noted. The third subset is then used to evaluate model performance, by calculating the *RMSE* value associated with the optimal n and D determined in the previous step. All possible combinations of three subsets for this optimization-evaluation procedure are employed, and an overall *RMSE* value is then computed as the root mean-square value of the three individual subset values (Eng and others, 2005; Eng and others, 2007a).

Use of the WREG Program

The remainder of this report provides details on how to use the Weighted-Multiple-Linear Regression Program (WREG). It can be used to set up, run, and evaluate a multiple-linear regression. As described in previous sections, the methods used by the program have been customized for use in the regionalization of streamflow characteristics. Many of the default methods may not be suitable for other regression problems.

The program is driven by a graphical user interface that leads the user through the process of setting up a regression. A number of input files are required, and detailed output is available in text files generated by WREG.

Notice.—MATLAB®. ©1984–2007 The MathWorks, Inc. was used to develop the graphical user interface and source code for WREG. The licensee's rights to deployment of WREG are governed by the license agreement between licensee and MathWorks, and licensee may not modify or remove any license agreement file that is included with the MCR libraries. Installation of the application denotes acceptance of the terms of the license specified in the file named license.txt in the folder WREGv1 included with this manual.

Program Requirements

- Windows operating system.
- Approximately 400 MB of free disk space.

Installation

1. Unzip the distribution file (WREGv1.zip) and extract it to the directory of your choosing. There should be four files—MCRInstaller.exe, WREGv1.exe, WREGv1.ctf, and license.txt—and two folders—WREG_Source_Code and Sample_Files.
2. Run the program MCRInstaller.exe. This program installs MATLAB Component Runtime, software that allows the WREG executable program to run. The installation

program leads you through the process with multiple GUI windows and may take several minutes to complete. Administrator privileges are required.

3. WREGv1.exe and WREGv1.ctf should be copied to another folder that contains input files and the program can be executed from that location. For example, copying the files into the folder Sample_Files will allow execution of the program using the sample input files. Alternatively, input files can be copied to the directory in which WREGv1.exe and WREGv1.ctf were installed.
4. WREGv1.exe is the program executable file and it is now ready to run. WREGv1.ctf is a required file that must be located in the same working directory as WREGv1. exe. After WREG is run for the first time, another folder, WREGv1_mcr, will be created in the working directory. It contains additional files for running WREG. These files should not be changed. (If a new version of WREG is provided, the files WREGv1.exe and WREGv1.ctf and the folder WREGv1_mcr should all be deleted from the working directory. The new WREG.exe and WREG.ctf should be placed in the directory. A new WREG_mcr folder will be created the first time WREG is executed.)
5. Double-click on WREGv1.exe to start WREG.

Input Files

The input files are listed in table 1 with a brief description. WREG automatically reads these files to set up the regression. Input files that are required for a regression analysis must be located in the same working directory as the WREGv1.exe executable file. As noted in table 1, some input files are always required for the program to run. Others are required only for certain WREG options, as noted in table 1. All input files are text files, and all fields within them are tab-delimited. The input files can be created, viewed, and edited in a spreadsheet program, such as Microsoft Excel, and then saved as a tab-delimited file. They can also be created, viewed, and edited using a text editor. Note that some text editors may not display the input files in an easy-to-read format, with properly aligned columns. A detailed description of each file's contents and format follows.

Table 1. WREG input files.

File name	Description	WREG requirements
SiteInfo.txt	Site information and basin characteristics to be used in the regression (the independent variables)	Always required.
FlowChar.txt	Flow characteristics to be used in the regression (the dependent variables)	Always required.
LP3G.txt	Skew for Log-Pearson Type III distribution	Always required.
LP3K.txt	K for Log-Pearson Type III distribution	Always required.
LP3s.txt	Standard deviation for Log-Pearson Type III distribution	Always required.
UserWLS.txt	User-specified weighting matrix.	Required only if the user-defined WLS option is selected.
USGS########.txt	Annual time series of flow at streamflow-gaging stations	Required only when using the GLS option. When needed, one file is required for each streamflow-gaging station listed in SiteInfo.txt.

The examples are based on the input files in the Sample_Files directory distributed with WREG and use peak-flow statistics and basin attributes for streamflow-gaging stations in Iowa.

SiteInfo.txt

This input file contains the independent variables that may be used in the regression. An example file is shown in figure 1. All fields are tab-delimited. The first row contains header information describing the contents of each column. Information about each streamflow-gaging station is listed in the rows below the header.

The first eight columns are required:

1. Station ID: The USGS identifier for each streamflow-gaging station.
2. Lat: Latitude of the streamflow-gaging station (or centroid of the watershed). The latitude and longitude are used to approximate the distance between streamflow-gaging stations in the GLS option. Latitude should be entered in decimal degrees (positive).
3. Long: Longitude of the streamflow-gaging station (or centroid of the watershed). Longitude should be entered in decimal degrees (either positive or negative).
4. No. Annual Series: The number of years of record available at the streamflow-gaging station, and used to estimate flow characteristics at the station. This information is used by WREG to assign weights to each streamflow gaging station. In general, only complete years of record should be used to generate flow characteristics.
5. Zero-1;NonZero-2: Flow characteristics (used for the dependent variable in a regression) will sometimes equal zero at a streamflow-gaging station. This field should show a 1 if the

dependent variable for a station is equal to zero, and 2 if not. This field is not currently used by the WREG, but a value must be provided. The field is intended to facilitate addition of logistic regression to future versions of WREG.

6. FreqZero: The number of instances in which the annual-time series values equals zero. For example, if the regression will estimate the $Q_{7,10}$, the number of years in which the 7-day minimum flow was equal to zero should be entered here.[1] Note that these values may need to change if a different flow statistic is used as the response variable. This field is not currently used by WREG, but a value must be provided. It is included to facilitate addition of logistic regression to future versions of WREG.

7. Regional Skew: This field is the value of the regional skew used to calculate a peak-flow frequency statistic. If a regional skew is not applicable, then a dummy value, such as -99.99, needs to be entered. This field is only used when forming a WLS or GLS regression, but a value must always be supplied.

8. Cont-1;PR-2: This field should show a 1 if the station is a continuous streamflow-gaging station, or a 2 if it is a partial-record site. For peak-flow studies, crest-stage gages should be treated like streamflow-gaging stations. In low-flow studies, partial-record sites (or miscellaneous sites) are sometimes used to supplement continuous-record stations. At the partial-record sites, only sporadic measurements are made. Estimation of flow characteristics at partial-record sites requires specialized techniques. The resulting estimates of flow characteristics at partial-record sites have larger uncertainty than estimates at continuous-record stations.

[1]The $Q_{7,10}$ is the lowest stream flow for 7 consecutive days that would be expected to occur on average once in 10 years. It is estimated using an annual-time series of the minimum 7-day flows.

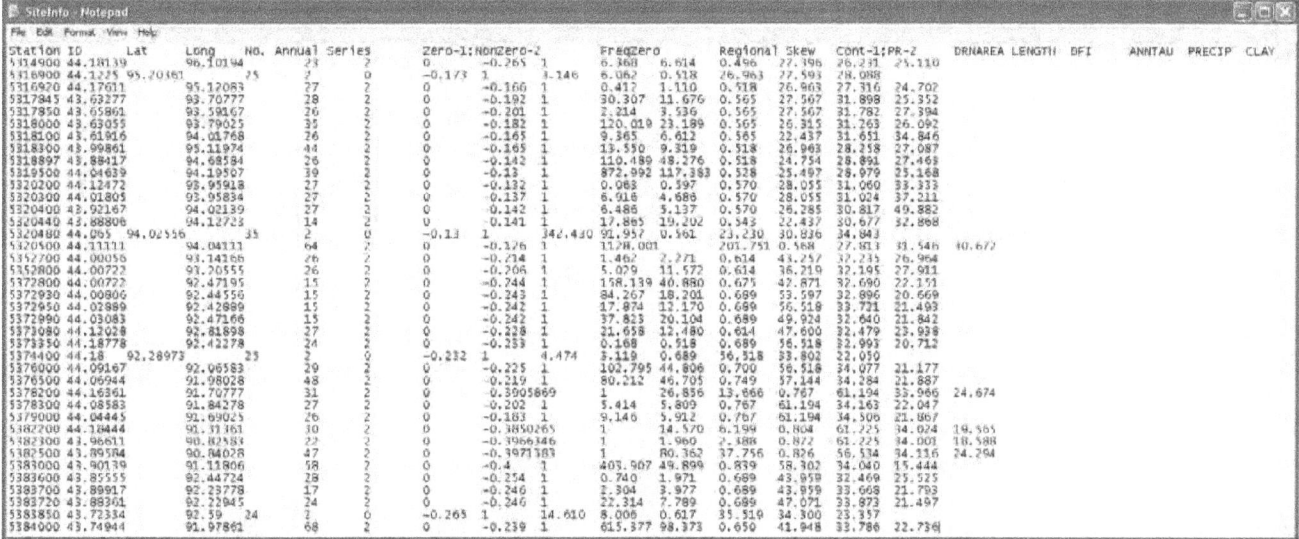

A

Figure 1.   Example of input file SiteInfo.txt in text tab-delimited format, as shown in *A,* the Notepad text editor and *B (next page),* Microsoft Excel.

SiteInfo [Compatibility Mode]

	A	B	C	D	E	F	G	H	I	J	K	L	M	N
1	Station ID	Lat	Long	No Annual Series	Zero-1;NonZero-2	FreqZero	Regional Skew	Cont-1;PR-2	DRNAREA	LENGTH	BFI	ANNTAU	PRECIP	CLAY
2	5314900	44.18139	96.10194	23	2	0	-0.265	1	6.368	6.614	0.496	27.396	26.231	25.11
3	5316900	44.1225	95.20361	25	2	0	-0.173	1	3.146	6.062	0.518	26.963	27.593	28.088
4	5316920	44.17611	95.12083	27	2	0	-0.166	1	0.412	1.11	0.518	26.963	27.316	24.702
5	5317845	43.63277	93.70777	28	2	0	-0.192	1	30.307	11.676	0.565	27.567	31.898	25.352
6	5317850	43.65861	93.59167	26	2	0	-0.201	1	2.214	3.536	0.565	27.567	31.782	27.394
7	5318000	43.63055	93.79025	35	2	0	-0.182	1	120.019	23.189	0.565	26.315	31.263	26.092
8	5318100	43.61916	94.01768	26	2	0	-0.165	1	9.365	6.612	0.565	22.437	31.651	34.846
9	5318300	43.99861	95.11974	44	2	0	-0.165	1	13.55	9.319	0.518	26.963	28.258	27.087
10	5318897	43.88417	94.68584	26	2	0	-0.142	1	110.489	48.276	0.518	24.754	28.891	27.463
11	5319500	44.04639	94.19507	39	2	0	-0.13	1	872.992	117.383	0.528	25.497	28.979	25.168
12	5320200	44.12472	93.95918	27	2	0	-0.132	1	0.063	0.597	0.57	28.055	31.06	33.333
13	5320300	44.01805	93.95834	27	2	0	-0.137	1	6.916	4.686	0.57	28.055	31.024	37.211
14	5320400	43.92167	94.02139	27	2	0	-0.142	1	6.486	5.137	0.57	26.285	30.817	49.882
15	5320440	43.88806	94.12723	14	2	0	-0.141	1	17.865	19.202	0.543	22.437	30.677	32.868
16	5320480	44.065	94.02556	35	2	0	-0.13	1	342.43	91.957	0.561	23.23	30.836	34.843
17	5320500	44.11111	94.04111	64	2	0	-0.126	1	1128.001	201.751	0.568	27.813	31.546	30.672
18	5352700	44.00056	93.14166	26	2	0	-0.214	1	1.462	2.271	0.614	43.257	32.235	26.964
19	5352800	44.00722	93.20555	26	2	0	-0.206	1	5.029	11.572	0.614	36.219	32.195	27.911
20	5372800	44.00722	92.47195	15	2	0	-0.244	1	158.139	40.88	0.675	42.871	32.69	22.151
21	5372930	44.00806	92.44556	15	2	0	-0.243	1	84.267	18.201	0.689	53.597	32.896	20.669
22	5372950	44.02889	92.42889	15	2	0	-0.242	1	17.874	12.17	0.689	56.518	33.721	21.493
23	5372990	44.03083	92.47166	15	2	0	-0.242	1	37.823	20.104	0.689	49.924	32.64	21.842
24	5373080	44.12028	92.81898	27	2	0	-0.228	1	21.658	12.48	0.614	47.6	32.479	23.938
25	5373350	44.18778	92.42278	24	2	0	-0.233	1	0.168	0.518	0.689	56.518	32.993	20.712
26	5374400	44.18	92.28973	25	2	0	-0.232	1	4.474	3.119	0.689	56.518	33.802	22.05
27	5376000	44.09167	92.06583	29	2	0	-0.225	1	102.795	44.806	0.7	56.518	34.077	21.177
28	5376500	44.06944	91.98028	48	2	0	-0.219	1	80.212	46.705	0.749	57.144	34.284	21.887
29	5378200	44.16361	91.70777	31	2	0	-0.3905869	1	26.856	13.666	0.767	61.194	33.966	24.674
30	5378300	44.08583	91.84278	27	2	0	-0.202	1	5.414	5.809	0.767	61.194	34.163	22.047
31	5379000	44.04445	91.69025	26	2	0	-0.183	1	9.146	5.912	0.767	61.194	34.506	21.867
32	5382200	44.18444	91.31361	30	2	0	-0.3850265	1	14.57	6.199	0.804	61.225	34.024	19.565
33	5382300	43.96611	90.82583	22	2	0	-0.3966346	1	1.96	2.388	0.872	61.225	34.001	18.588
34	5382500	43.89584	90.84028	47	2	0	-0.3971383	1	80.362	37.756	0.826	56.534	34.116	24.294
35	5383000	43.90139	91.11806	58	2	0	-0.4	1	403.907	49.899	0.839	58.302	34.04	15.444
36	5383600	43.85555	92.44724	28	2	0	-0.254	1	0.74	1.971	0.689	43.959	32.469	25.525
37	5383700	43.89917	92.23778	17	2	0	-0.246	1	2.304	3.977	0.689	43.959	33.668	21.793
38	5383720	43.88361	92.22945	24	2	0	-0.246	1	22.314	7.789	0.689	47.071	33.873	21.497
39	5383850	43.72334	92.59	24	2	0	-0.265	1	14.61	8.006	0.617	35.519	34.3	23.357
40	5384000	43.74944	91.97861	68	2	0	-0.239	1	615.377	98.373	0.65	41.948	33.786	22.736

SiteInfo

B

Figure 1. (continued)

9. In the columns following these required entries, independent variables (basin characteristics) should be entered. The headers should be descriptive, as they will be displayed by WREG for selection of independent variables. For display purposes, a maximum of 11 characters is recommended.

Streamflow-gaging stations must be listed in ascending alphanumeric order (by USGS station identifier numbers) in this file and all others in which each row contains information for one station (FlowChar.txt, LP3G.txt, LP3K.txt, LP3s.txt). The order in which the gaging stations are listed should be the same as the order in which the USGS########.txt files are shown in Windows Explorer when sorted by name.

FlowChar.txt

This file contains the dependent variables that may be selected for use in the regression. An example is shown in figure 2. All fields are tab-delimited. The first row contains header information describing the contents of each column. The first column, Station ID, is required and entries should be identical to those in column 1 of the SiteInfo.txt file. Note that the program requires all streamflow-gaging stations to be entered in the same order in both files.

Following the Station ID column, the remaining columns are used to define flow characteristics. The names used in the header should be descriptive, as they will later be used by the WREG to solicit user input. For display purposes, a maximum of 11 characters is recommended.

LP3G.txt

This file contains the skew value for each streamflow-gaging station that was used when fitting the log-Pearson Type III distribution. An example is shown in figure 3. All fields are tab-delimited. The first row contains header information describing the contents of each column. The first column, Station ID, is

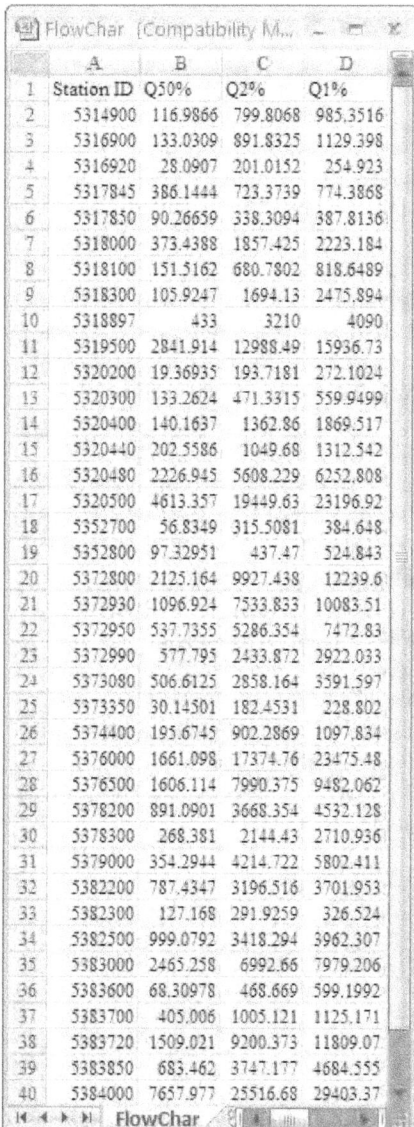

	A	B	C	D
1	Station ID	Q50%	Q2%	Q1%
2	5314900	116.9866	799.8068	985.3516
3	5316900	133.0309	891.8325	1129.398
4	5316920	28.0907	201.0152	254.923
5	5317845	386.1444	723.3739	774.3868
6	5317850	90.26659	338.3094	387.8136
7	5318000	373.4388	1857.425	2223.184
8	5318100	151.5162	680.7802	818.6489
9	5318300	105.9247	1694.13	2475.894
10	5318897	433	3210	4090
11	5319500	2841.914	12988.49	15936.73
12	5320200	19.36935	193.7181	272.1024
13	5320300	133.2624	471.3315	559.9499
14	5320400	140.1637	1362.86	1869.517
15	5320440	202.5586	1049.68	1312.542
16	5320480	2226.945	5608.229	6252.808
17	5320500	4613.357	19449.63	23196.92
18	5352700	56.8349	315.5081	384.648
19	5352800	97.32951	437.47	524.843
20	5372800	2125.164	9927.438	12239.6
21	5372930	1096.924	7533.833	10083.51
22	5372950	537.7355	5286.354	7472.83
23	5372990	577.795	2433.872	2922.033
24	5373080	506.6125	2858.164	3591.597
25	5373350	30.14501	182.4531	228.802
26	5374400	195.6745	902.2869	1097.834
27	5376000	1661.098	17374.76	23475.48
28	5376500	1606.114	7990.375	9482.062
29	5378200	891.0901	3668.354	4532.128
30	5378300	268.381	2144.43	2710.936
31	5379000	354.2944	4214.722	5802.411
32	5382200	787.4347	3196.516	3701.953
33	5382300	127.168	291.9259	326.524
34	5382500	999.0792	3418.294	3962.307
35	5383000	2465.258	6992.66	7979.206
36	5383600	68.30978	468.669	599.1992
37	5383700	405.006	1005.121	1125.171
38	5383720	1509.021	9200.373	11809.07
39	5383850	683.462	3747.177	4684.555
40	5384000	7657.977	25516.68	29403.37

Figure 2. Example of input file FlowChar.txt in text tab-delimited format, as shown in Microsoft Excel.

required and entries must match those in column 1 of the SiteInfo.txt and FlowChar.txt files. Again, the program requires all streamflow-gaging stations to be entered in the same order in both files, in ascending alphanumeric gage site order.

Following the Station ID column, skew values for each of the frequency statistics listed in FlowChar.txt should be entered. These columns should correspond to those used in FlowChar.txt (for example, if column 2 of FlowChar.txt contains a peak for statistic, the skew value in column 2 of LP3G.txt should be

	A	B	C	D
1	Station ID	Skew50%	Skew2%	Skew1%
2	5314900	-0.4378513	-0.4378513	-0.4378513
3	5316900	-0.1628717	-0.1628717	-0.1628717
4	5316920	-0.224812	-0.224812	-0.224812
5	5317845	-0.4373209	-0.4373209	-0.4373209
6	5317850	-0.5238652	-0.5238652	-0.5238652
7	5318000	-0.3778404	-0.3778404	-0.3778404
8	5318100	-0.1882706	-0.1882706	-0.1882706
9	5318300	0.0813257	0.0813257	0.0813257
10	5318897	-0.2199	-0.2199	-0.2199
11	5319500	0.0368123	0.0368123	0.0368123
12	5320200	0.301714	0.301714	0.301714
13	5320300	0.0716433	0.0716433	0.0716433
14	5320400	0.123498	0.123498	0.123498
15	5320440	0.0607876	0.0607876	0.0607876
16	5320480	-0.2773506	-0.2773506	-0.2773506
17	5320500	-0.1934676	-0.1934676	-0.1934676
18	5352700	-0.3161195	-0.3161195	-0.3161195
19	5352800	-0.2169663	-0.2169663	-0.2169663
20	5372800	0.060633	0.060633	0.060633
21	5372930	0.382059	0.382059	0.382059
22	5372950	0.385835	0.385835	0.385835
23	5372990	-0.1072303	-0.1072303	-0.1072303
24	5373080	-0.01383939	-0.01383939	-0.01383939
25	5373350	-0.1332477	-0.1332477	-0.1332477
26	5374400	-0.08422817	-0.08422817	-0.08422817
27	5376000	-0.08696695	-0.08696695	-0.08696695
28	5376500	-0.4687821	-0.4687821	-0.4687821
29	5378200	0.341917	0.341917	0.341917
30	5378300	-0.3649173	-0.3649173	-0.3649173
31	5379000	-0.06969751	-0.06969751	-0.06969751
32	5382200	-0.5004891	-0.5004891	-0.5004891
33	5382300	0.0400738	0.0400738	0.0400738
34	5382500	-0.2367817	-0.2367817	-0.2367817
35	5383000	-0.117111	-0.117111	-0.117111
36	5383600	-0.09854396	-0.09854396	-0.09854396
37	5383700	-0.1627242	-0.1627242	-0.1627242
38	5383720	0.105583	0.105583	0.105583
39	5383850	-0.02941165	-0.02941165	-0.02941165
40	5384000	-0.2773225	-0.2773225	-0.2773225

Figure 3. Example of input file LP3G.txt as shown in Microsoft Excel. Note that the skew values entered for each streamflow gage are the same, since each flow statistic (Q2, Q50, and Q100) is based on the same annual-peak-flow series.

that for the annual peak flowtime series). Each statistic in FlowChar.txt must have a corresponding skew value. Note that if each statistic is based on the same annual series, the skews will be identical. If the statistic is not a frequency statistic and no skew was used in its computation, a dummy value of -99.99 may be used in the LP3G.txt file. For peak-flow studies, the weighted skew (equation 16) should be entered in LP3G.txt if it the weighted skew was used when fitting the log-Pearson Type III distribution. For other flow characteristics where a regional skew is not calculated, the at-site skew values, g, should be used. Information in this file is used by the WLS and GLS options.

LP3K.txt

This file contains the log-Pearson Type III distribution standard deviate, K, values (K values for a given exceedance probability and skew are tabulated in Bulletin 17B). An example is shown in figure 4. All fields are tab-delimited. The first row contains header information describing the contents of each column. The first column, Station ID, is required and must match the entries in column 1 of the SiteInfo.txt and FlowChar.txt files. Again, the program requires all streamflow-gaging stations to be entered in the same order in both files.

Following the Station ID column, K values for each of the frequency statistics listed in FlowChar.txt should be entered. These columns

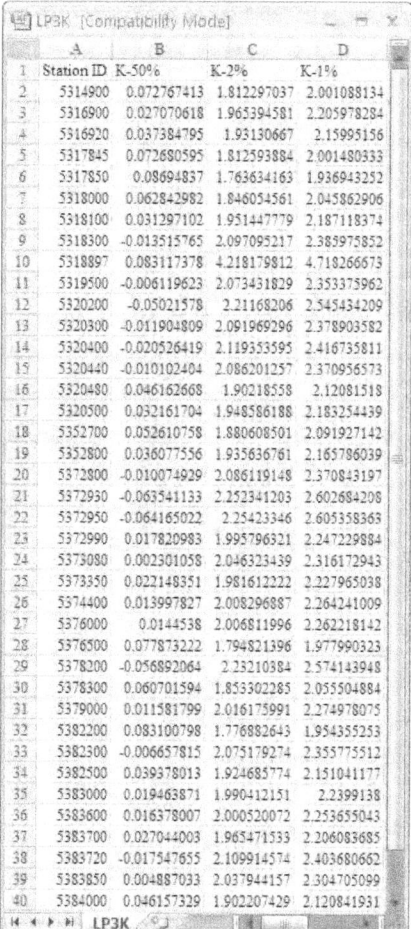

	A	B	C	D
1	Station ID	K-50%	K-2%	K-1%
2	5314900	0.072767413	1.812297037	2.001088134
3	5316900	0.027070618	1.965394581	2.205978284
4	5316920	0.037384795	1.93130667	2.15995156
5	5317845	0.072680595	1.812593884	2.001480333
6	5317850	0.08694837	1.763634163	1.936943252
7	5318000	0.062842982	1.846054561	2.045862906
8	5318100	0.031297102	1.951447779	2.187118374
9	5318300	-0.013515765	2.097095217	2.385975852
10	5318897	0.083117378	4.218179812	4.718266673
11	5319500	-0.006119623	2.073431829	2.353375962
12	5320200	-0.05021578	2.21168206	2.545434209
13	5320300	-0.011904809	2.091969296	2.378903582
14	5320400	-0.020526419	2.119353595	2.416735811
15	5320440	-0.010102404	2.086201257	2.370956573
16	5320480	0.046162668	1.90218558	2.12081518
17	5320500	0.032161704	1.948586188	2.183254439
18	5352700	0.052610758	1.880608501	2.091927142
19	5352800	0.036077556	1.935636761	2.165786039
20	5372800	-0.010074929	2.086119148	2.370843197
21	5372930	-0.063541133	2.252341203	2.602684208
22	5372950	-0.064165022	2.25423346	2.605358363
23	5372990	0.017820983	1.995796321	2.247229884
24	5373080	0.002301058	2.046323439	2.316172943
25	5373350	0.022148351	1.981612222	2.227965038
26	5374400	0.013997827	2.008296887	2.264241009
27	5376000	0.0144538	2.006811996	2.262218142
28	5376500	0.078873222	1.794821396	1.977990323
29	5378200	-0.056892064	2.23210384	2.574143948
30	5378300	0.060701594	1.853302285	2.055504884
31	5379000	0.011581799	2.016175991	2.274978075
32	5382200	0.083100798	1.776882643	1.954355253
33	5382300	-0.006657815	2.075179274	2.355775512
34	5382500	0.039378013	1.924685774	2.151041177
35	5383000	0.019463871	1.990412151	2.2399138
36	5383600	0.016378007	2.000520072	2.253655043
37	5383700	0.027044003	1.965471533	2.206083685
38	5383720	-0.017547655	2.109914574	2.403680662
39	5383850	0.004887033	2.037944157	2.304705099
40	5384000	0.046157329	1.902207429	2.120841931

Figure 4. Example of input file LP3K.txt as shown in Microsoft Excel.

should correspond to those used in FlowChar.txt (that is, if column 2 of FlowChar.txt contains the $Q_{7,10}$, the K value in column 2 of LP3K.txt should be that for the 7-day annual-time series). Each statistic in FlowChar.txt must have a corresponding K value. If the statistic is not a frequency statistic, a dummy value of -99.99 may be used in the LP3K.txt file. Information in this file is used by the WLS and GLS options.

LP3s.txt

This file contains the standard deviation, σ, of the annual-time series that was fit to the log-Pearson Type III

	A	B	C	D
1	Station ID	s-50%	s-2%	s-1%
2	5314900	0.479928	0.479928	0.479928
3	5316900	0.426312	0.426312	0.426312
4	5316920	0.451268	0.451268	0.451268
5	5317845	0.156682	0.156682	0.156682
6	5317850	0.342215	0.342215	0.342215
7	5318000	0.390695	0.390695	0.390695
8	5318100	0.339842	0.339842	0.339842
9	5318300	0.570427	0.570427	0.570427
10	5318897	0.2104	0.2104	0.2104
11	5319500	0.317351	0.317351	0.317351
12	5320200	0.442131	0.442131	0.442131
13	5320300	0.260766	0.260766	0.260766
14	5320400	0.461622	0.461622	0.461622
15	5320440	0.340841	0.340841	0.340841
16	5320480	0.216116	0.216116	0.216116
17	5320500	0.326073	0.326073	0.326073
18	5352700	0.407219	0.407219	0.407219
19	5352800	0.343608	0.343608	0.343608
20	5372800	0.319362	0.319362	0.319362
21	5372930	0.361348	0.361348	0.361348
22	5372950	0.428135	0.428135	0.428135
23	5372990	0.315739	0.315739	0.315739
24	5373080	0.367614	0.367614	0.367614
25	5373350	0.399056	0.399056	0.399056
26	5374400	0.332854	0.332854	0.332854
27	5376000	0.511717	0.511717	0.511717
28	5376500	0.405831	0.405831	0.405831
29	5378200	0.26848	0.26848	0.26848
30	5378300	0.503492	0.503492	0.503492
31	5379000	0.53647	0.53647	0.53647
32	5382200	0.359233	0.359233	0.359233
33	5382300	0.173354	0.173354	0.173354
34	5382500	0.283354	0.283354	0.283354
35	5383000	0.229727	0.229727	0.229727
36	5383600	0.421534	0.421534	0.421534
37	5383700	0.203648	0.203648	0.203648
38	5383720	0.369036	0.369036	0.369036
39	5383850	0.363487	0.363487	0.363487
40	5384000	0.281625	0.281625	0.281625

Figure 5. Example of input file LP3s.txt as shown in Microsoft Excel.

distribution. An example is shown in figure 5. All fields are tab-delimited. The first row contains header information describing the contents of each column. The first column, Station ID, is required and must match the entries in column 1 of the SiteInfo.txt and FlowChar.txt files. Again, the program requires all streamflow-gaging stations to be entered in the same order in both files.

Following the Station ID column, σ values for each of the frequency statistics listed in FlowChar.txt should be entered. These columns should correspond to those used in FlowChar.txt (that is, if column 2 of FlowChar.txt contains the $Q_{7,10}$, the σ value in column 2 of LP3K.txt should be that for the 7-day annual-time series). Each statistic in FlowChar.txt must have a corresponding value for σ. If the statistic is not a frequency statistic and σ is not used in its computation, a dummy value of -99.99 may be used in the LP3s.txt file. Information in this file is used by the WLS and GLS options.

UserWLS.txt

A user-defined weighting matrix can be specified in UserWLS.txt. This file contains no header information and simply contains the desired values of the Λ matrix (for example, equations 10, 12, and 19). The Λ matrix is a square matrix containing n rows and n columns, where n is the number of sites used in the regression analysis. WREG inverts this matrix to assign weights to each station used in the analysis. When using WLS, only the main diagonal of this matrix contains nonzero values. In the UserWLS.txt file, observations that have smaller variance and are thus considered to be more reliable should be given smaller values. When the matrix is inverted by WREG, these observations will be given larger weight. Observations with large variance (for example, at partial-record sites) would be considered less reliable and should be assigned larger values. When the matrix is inverted by WREG, these observations will be given smaller weight. For a GLS regression,

Figure 6. Example of input file UserWLS.txt as shown in Microsoft Excel. Because of the large number of columns in this file, it is difficult to read in a text editor.

matrix elements off the main diagonal may contain nonzero values. Figure 6 shows a portion of a UserWLS.txt file for a WLS regression problem.

USGS########.txt

These files contain the annual-time series of interest for each streamflow-gaging station used in the analysis. They are only used when the GLS option is selected. The Station ID should be substituted in the filename for the #s. WREG looks for any file with "USGS" in the filename to use as input files. Consequently, no other files in the directory should have a filename containing USGS. WREG reads in each file sequentially in alphanumeric ascending order.[2] An example file is shown in figure 7.

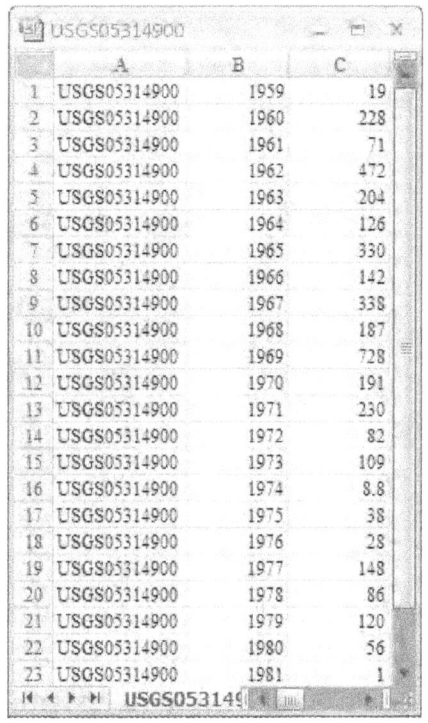

Figure 7. An example of a USGS#######.txt file for streamflow-gaging station 09183000.

---

[2]Other characters can be added to the filename, but they should be identical for each set of USGS#######.txt files. For example, a set of peak flow files could follow the naming convention USGS#######.peak.txt while a set of low-flow files could follow the naming convention USGS#######.lowflow.txt. Only one set of these files can be in the working directory at one time.

Each file should include three columns and m rows, where m is the number of years of record available at the streamflow-gaging station. The first column contains the site identifier Station ID that was used in the other input files. The second column contains the year, and the third column contains the value of the flow statistic during that year. For example, for a regression to estimate the 1% chance exceedance flood (Q100, the 100-year return period flood), the annual-flow statistic entered in this file should be the annual-peak flow. The year may be a calendar year, water year, or climatic year, depending on what was used to define the annual series.

Note that even if more than one type of statistic is specified in the FlowChar.txt file, only one annual-time series can be entered in the USGS#######.txt file. Consequently, the USGS#######.txt files must be updated if a second time series is used to define the dependent variable. (In other words, the same USGS#######.txt files will not be valid for both peak flows and low flows.)

Running the Program

The WREG program uses a series of windows to guide the user through the formation, selection, and evaluation of a regression. The DOS window is displayed upon program execution, and will occasionally print information that monitors the progress of the program. Important information is always printed to a GUI window or to an output file.

Run the program by double-clicking on WREGv1.exe. If required input files are missing, the program will not run.

Set Up Model

Select Variables

The first window (fig. 8) is used to select variables (independent and dependent) for the regression. The items shown in the dependent variables menu

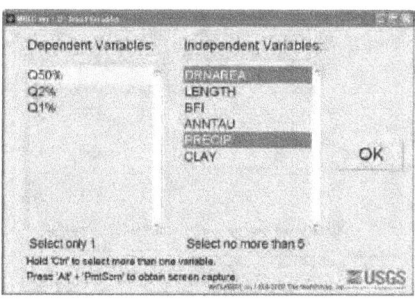

Figure 8. Example of WREG window used to select variables to be used in the regression.

and in the independent variables menu will depend on what has been included in the input files. One dependent variable should be selected. This is the flow statistic that will be estimated by the regression. To select more than one independent variable, hold down the CTRL key while clicking the left mouse button. Up to five independent variables can be selected.

Select Transformations

Each variable can be transformed by WREG (equation 5). Figure 9 shows the GUI window that facilitates variable transformation. The dependent variable and each independent variable are listed on the left of the window. In the middle part of this window, radio buttons are used to select whether to apply no transformations, to transform the variable using logarithms (base 10) (log10[…]), using natural logarithms (ln[…]), or using an exponential function (e[…]). By default, no transformations are applied. In the right part of the window, additional transformations can be applied to the variable. The logarithmic or exponential transformation will be applied to the entire expression specified in the right half of the window. The default values that are displayed initially will result in no additional changes to the variable. C1 is used to multiply the variable by the constant C1. C2 is used to raise the variable to the power C2. C3 is used to add a constant C3 to the variable. C4 is used to raise the variable (including any algebraic transformation specified by C1, C2, or C3) to a power C4.

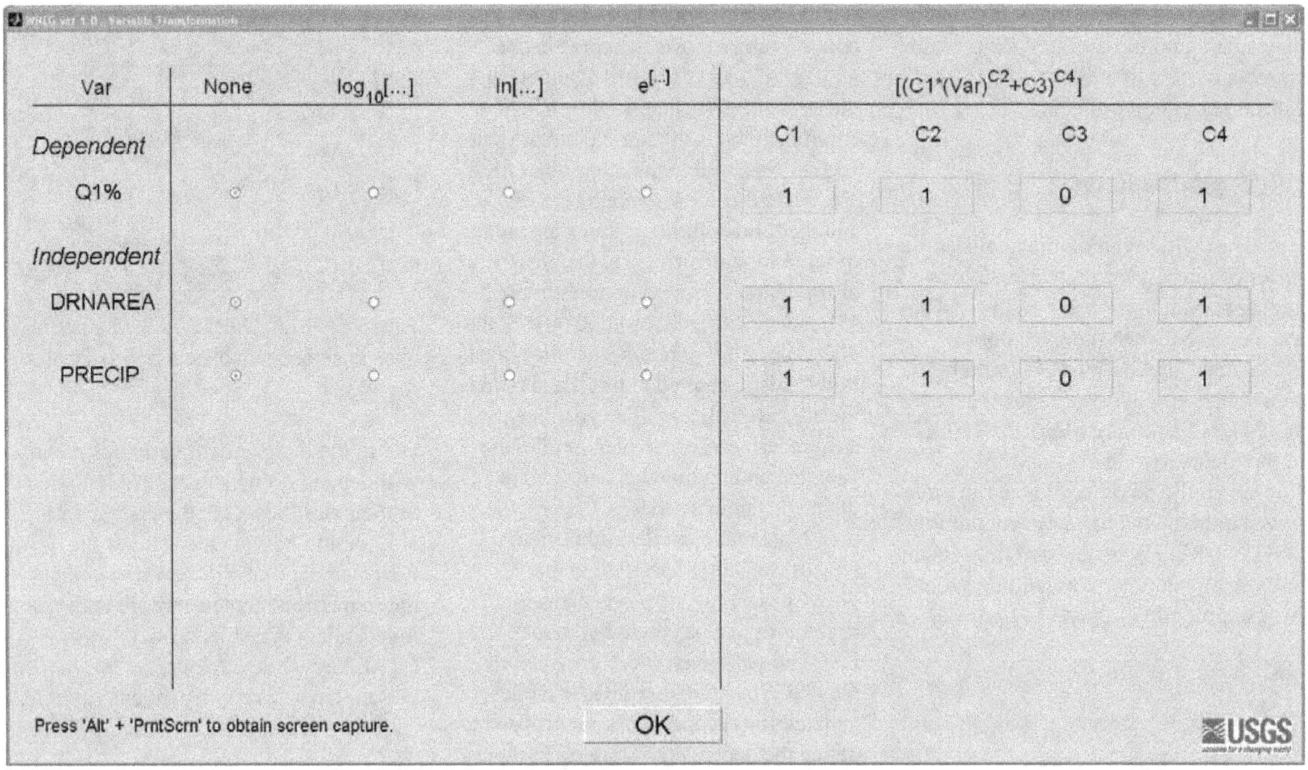

A

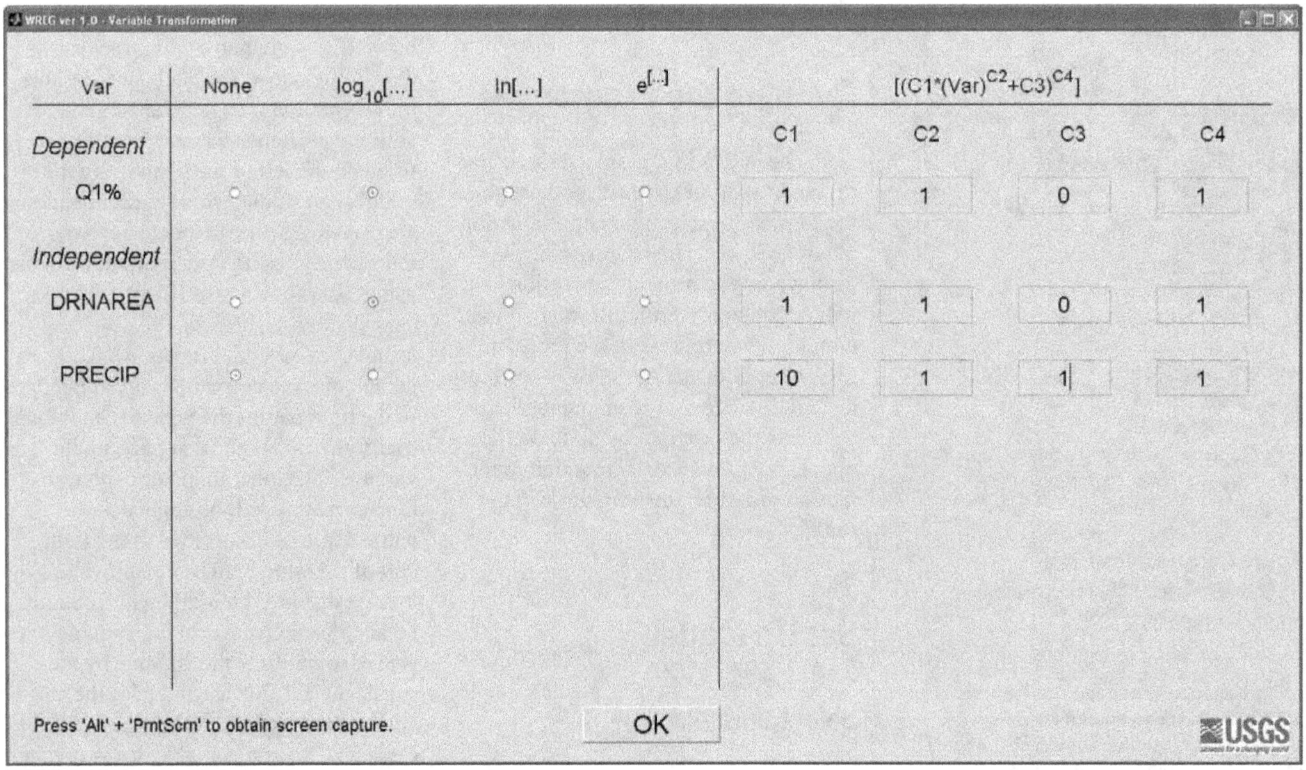

B

Figure 9.   *A*, Example of WREG window for selecting transformations, as it initially appears. *B*, After transformations have been specified.

For example, in figure 9A, the initial GUI window is shown, with no transformations selected. In figure 9B, selections will cause 1% chance exceedance flood (Q1%) and the drainage area (DRNAREA) to be log (base 10) transformed. No further transformations will be applied to these variables. The annual precipitation (PRECIP) will first be multiplied by 10 and then a constant of 1 will be added to that quantity. These are merely illustrations of the use of the transformations, and are not intended to suggest useful transformations for these variables.

Model Selection

Either conventional regression or region-of-influence (RoI) regression can be selected by the radio buttons in the upper portion of the GUI window shown in figure 10. Equations formed by conventional regression can be used to estimate statistics at ungaged locations. WREG forms a RoI regression for each streamflow-gaging station in the database, allowing for evaluation of RoI as a regression strategy. However, WREG does not provide a function allowing the user to form a RoI regression for an ungaged location.

For RoI, the method for defining the region of influence must also be selected—either based on geographic proximity (GRoI), proximity in predictor variable space (PRoI), or a hybrid approach combining geographic and predictor variable space (HRoI). (These methods were discussed in the section **Definition of Regions**.) If using RoI, the number of sites, *n*, to use in the region must also be specified in the field under the RoI radio buttons. If the GRoI or HRoI method is selected, then the geographic proximity, *D*, must also be specified; this

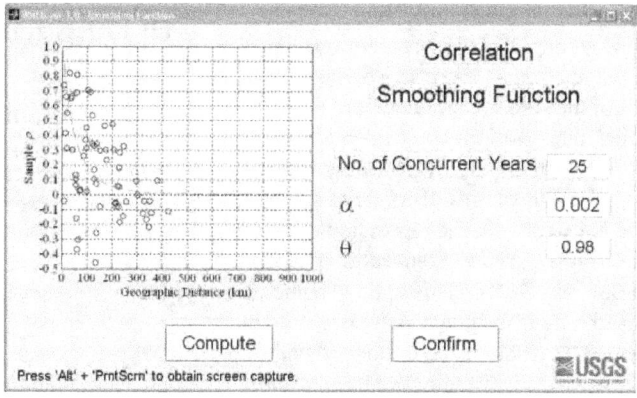

Figure 11. Example of WREG window for selecting parameters of the smoothing function for correlation as a function of distance between streamflow-gaging stations.

is the maximum distance from the ungaged location from which sites can be selected to form a region.

In the bottom panel of the same window (fig. 10), the parameter estimation method is selected using radio buttons as either ordinary least squares (OLS), weighted least squares (WLS), or generalized least squares (GLS).

If WLS or GLS is chosen, the program uses methods described earlier in this report to calculate the weighting matrix, Λ. The USGS########.txt files need to be supplied as input files if the GLS option is selected.

If conventional regression methods are used, there is also an option for utilizing user-specified weights in the regression. If UserWLS.txt was not supplied as an input file, an error message will appear if User Specified Weights is selected.

If a GLS regression is selected, a screen appears that allows adjustment of the parameters of a smoothing function (equation 18) that relates the correlation between streamflow time series (input in USGS########.txt files) at two sites to the geographic distance between them, as shown in figure 11. The default settings are a good starting point, but values can be adjusted to better fit specific datasets.

The "No. of Concurrent Years" specifies the minimum overlap among annual-peak flow records at streamflow-gaging stations before the correlation between them is used to estimate the smoothing function. If too few concurrent years are specified, the variability in correlation among streamflow-gaging stations may be too large to discern a relation between correlation and distance. However, if the required number of concurrent years is too large, there may not be enough data points to adequately define the relation. Generally, to fit well, the value of alpha, α, will be positive and near zero while theta, θ, will be slightly less than 1. This combination of parameters causes the correlation to be near 1 when streamflow-gaging stations are very close, and to asymptotically approach a value near zero as the distance between them increases. When values are in this range, the closer θ is to one, the more slowly the correlation will decrease with distance. The closer α is to zero, the closer the asymptote will lie to zero.

After making desired changes to the parameters, click the "Compute" button to (re)compute and draw the smoothing

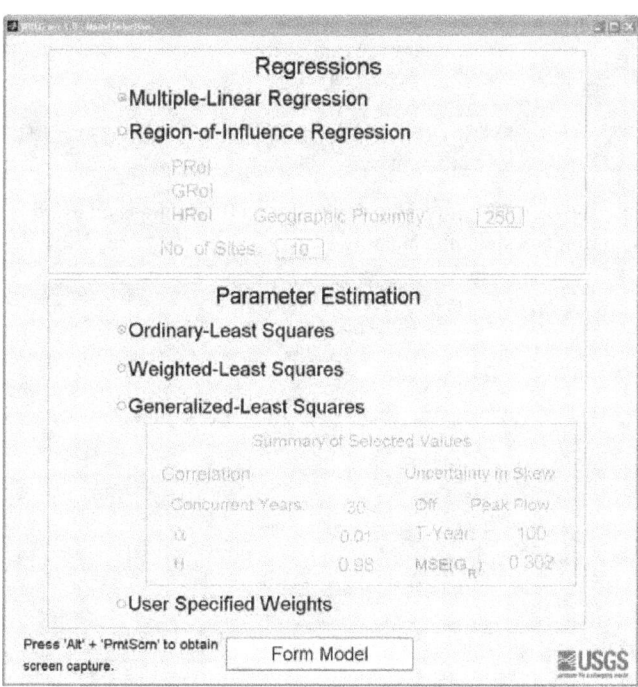

Figure 10. Example of WREG window for selecting the regression.

function. The "Confirm" button will be grayed out while the program is computing the change. When satisfied with the fit of the smoothing function to the data, click the "Confirm" button.

Another window (fig. 12) will pop up that allows the uncertainty in skew to be considered in the regression calculations. [See section **Generalized Least Squares (GLS)**.] This option should be used only if weighted skew was used in the calculation of the frequency statistics. The radio buttons are used to specify whether peak flows or low flows are being considered. Then the recurrence interval of the frequency statistic under consideration should be entered under T-Year, and the mean square error of the regional skew should be entered under $MSE(G_R)$. A default value of 0.302 is used in the program because this value is associated with plate I of Bulletin 17B of the Interagency Advisory Committee on Water Data (1982).

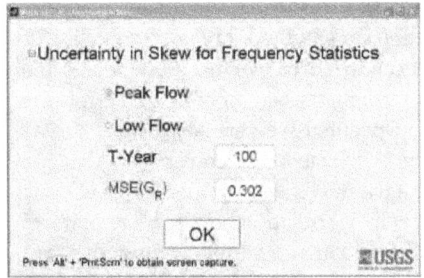

Figure 12. Example of WREG window for selecting option to include uncertainty in skew.

GUI Outputs

Regression Summary

Once parameters for the regression are calculated, the regression summary window (fig. 13) displays some basic performance metrics (appropriate to the type of regression used) and the regression equation. Elements of the regression equation that are shown in red are not statistically significant at the 5% level. Three plots accompany the regression output and are shown in figure 14. More detailed regression

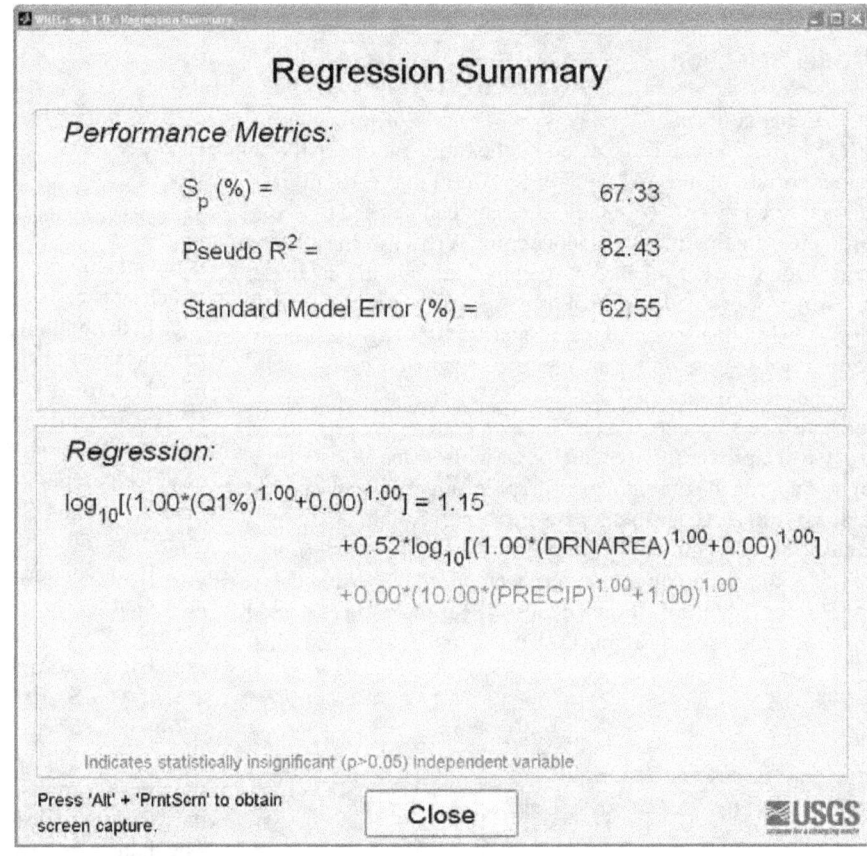

Figure 13. WREG window showing the regression results for an GLS regression using the parameters and transformations specified in figure 9.

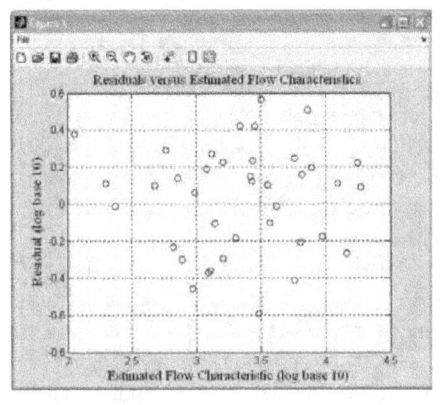

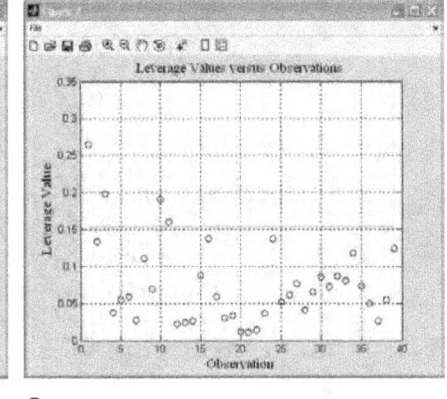

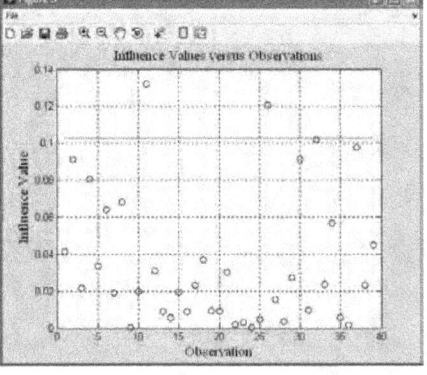

A   B   C

Figure 14. *A*, Example of plot showing residuals versus estimated flow characteristics. *B*, Example of plot showing leverage values versus observations. *C*, Example of plot showing influence values versus observations.

results, including p-values for individual parameters, are available in the text output files.

Residuals Versus Estimated Flow Characteristics

This plot (fig. 14A) shows the regression residuals (equation 28) plotted against the estimated flow characteristics. Both axes will be in units of the transformed flow characteristics [for example, log (base 10)].

Leverage Values Versus Observations

This plot (figure 14B) shows the leverage values plotted by observation number. Observations are numbered in the order in which they appear in the input files. The red line shows the threshold (as calculated by equation 41) above which an observation may be considered to have high leverage, as calculated by the appropriate equation for the type of regression, discussed earlier in the section **Leverage and Influence Statistic**. Observations with high leverage should be checked carefully for possible errors.

Influence Values Versus Observations

This plot (fig. 14C) shows influence values plotted by observation number. The red line shows the threshold (as calculated by equation 45) above which an observation may be considered to have large influence, as calculated by the appropriate equation for the type of regression, discussed earlier in the section **Leverage and Influence Statistic**. Observations with high influence should be checked carefully for possible errors.

Output Files

The output files are listed in table 2 with a brief description. A detailed description of each file's contents follows. A text editor can be used to view all of the output files. Each file can also be imported into a spreadsheet program, such as Microsoft Excel, as a text, tab-delimited file.

```
Regression Model for Q1%

Performance Metrics
Mean Squared Error               0.085
R2                               77.98
Standard Model Error             NA
*Note: R2 is in percent.

Coefficients of Model
                       Std Error   T value    P>|T|
Constant       0.97        0.70      1.39     0.173
DRNAREA        0.51        0.05     10.63     0.000
PRECIP         0.01        0.00      2.54     0.015

Transformations           C1        C2        C3        C4
Q1%           log10      1.00      1.00      0.00      1.00
DRNAREA       log10      1.00      1.00      0.00      1.00
PRECIP            1     10.00      1.00      1.00      1.00

Residuals, Leverage, and Influence of Observations
Leverage Limit                  0.154
Influence Limit                 0.103
       Observation   Residual  Leverage  Influence
                 1     0.145     0.216      0.002
                 2     0.286     0.143      0.005
                 3     0.108     0.200      0.001
                 4     0.622     0.028      0.004
                 5     0.333     0.045      0.002
                 6     0.435     0.053      0.004
                 7     0.322     0.028      0.001
                 8     0.265     0.107      0.003
                 9     0.020     0.110      0.000
                10     0.105     0.179      0.001
                11     0.346     0.183      0.011
                12     0.385     0.034      0.002
                13     0.165     0.037      0.000
                14     0.207     0.037      0.001
                15     0.196     0.088      0.001
                16     0.069     0.125      0.000
                17     0.269     0.056      0.002
                18     0.407     0.033      0.002
                19     0.165     0.054      0.001
                20     0.210     0.042      0.001
                 .
                 .
                 .
```

Figure 15. Example of output file ConventionalOLS.txt. Residuals, leverage, and influence are shown only for the first 20 observations.

Table 2. WREG output files.

File name	Description
ConventionalGLS.txt	Results from a GLS regression.
ConventionalOLS.txt	Results from an OLS regression.
ConventionalWLS.txt	Results from a WLS regression.
RegionofInfluenceOLS.txt	Results from an OLS regression using RoI.
RegionofInfluenceWLS.txt	Results from a WLS regression using RoI.
RegionofInfluenceGLS.txt	Results from a GLS regression using RoI.
RegressionModel.txt	The regression equation (including transformations) calculated by WREG. Output only for conventional regression.
InvXLX.txt	Covariance of the regression parameters (equation 46). Output only for conventional regression.
SSres.txt	Residual sum of squares (equation 36). Output only for conventional regression.
SStot.txt	Total sum of squares (equation 37). Output only for conventional regression.
EventLog.txt	Record of the program's execution. Helpful for diagnosing runtime errors.

ConventionalOLS.txt, ConventionalWLS.txt, and ConventionalGLS.txt

These three files provide information on conventional regressions and are the same except for the performance metrics that are displayed at the head of the file. The metrics shown are appropriate to the parameter selection scheme: OLS, WLS, or GLS. Figure 15 shows an example of output in ConventionalOLS.txt, figure 16 shows an example of output in ConventionalWLS.txt, and figure 17 shows an example of output in ConventionalGLS.txt.

```
Regression Model for Q1%

Performance Metrics
Avg Standard Error of Prediction      69.33
Pseudo R2                             80.92
Standard Model Error                  65.67
*Note: Avg Standard Error of Prediction, Pseudo R2,
and Standard Model Error are in percent.

Coefficients of Model
                    Std Error   T value   P>|T|
Constant     1.01     0.70       1.45     0.157
DRNAREA      0.51     0.05      10.77     0.000
PRECIP       0.01     0.00       2.51     0.017

Transformations          C1       C2       C3       C4
Q1%          log10     1.00     1.00     0.00     1.00
DRNAREA      log10     1.00     1.00     0.00     1.00
PRECIP          1     10.00     1.00     1.00     1.00

Residuals, Leverage, and Influence of Observations
Leverage Limit                        0.154
Influence Limit                       0.103
         Observation  Residual  Leverage  Influence
                  1    0.140     0.208     0.025
                  2    0.281     0.141     0.059
                  3    0.101     0.200     0.013
                  4    0.620     0.028     0.046
                  5    0.333     0.046     0.022
                  6    0.433     0.054     0.047
                  7    0.322     0.028     0.012
                  8    0.262     0.115     0.043
                  9    0.021     0.107     0.000
                 10    0.107     0.185     0.013
                 11    0.342     0.182     0.127
                 12    0.385     0.035     0.022
                 13    0.164     0.038     0.004
                 14    0.207     0.031     0.005
                 15    0.193     0.089     0.017
                 16    0.073     0.135     0.004
                 17    0.269     0.056     0.018
                 18    0.406     0.033     0.023
                 19    0.169     0.045     0.005
                 20    0.213     0.036     0.006
                 21    0.382     0.035     0.019
                 22    0.133     0.026     0.002
                 23    0.089     0.027     0.001
                 24    0.056     0.140     0.002
                 25    0.148     0.053     0.005
                 26    0.472     0.061     0.063
                 27    0.122     0.067     0.005
                 28    0.061     0.047     0.001
                 29    0.183     0.060     0.009
                 30    0.378     0.062     0.040
                 31    0.106     0.049     0.002
                 32    0.502     0.072     0.080
                 33    0.248     0.063     0.019
                 34    0.299     0.102     0.050
                 35    0.062     0.078     0.001
                 36    0.018     0.057     0.000
                 37    0.524     0.043     0.050
                 38    0.193     0.053     0.009
                 39    0.188     0.114     0.023
```

Figure 16.   Example of output file ConventionalWLS.txt.

```
Regression Model for Q1%

Performance Metrics
Avg Standard Error of Prediction      67.33
Pseudo R2                             82.43
Standard Model Error                  62.55
*Note: Avg Standard Error of Prediction, Pseudo R2,
and Standard Model Error are in percent.

Coefficients of Model
                    Std Error   T value   P>|T|
Constant     1.15     0.78       1.49     0.146
DRNAREA      0.52     0.05      10.94     0.000
PRECIP       0.00     0.00       2.02     0.051

Transformations          C1       C2       C3       C4
Q1%          log10     1.00     1.00     0.00     1.00
DRNAREA      log10     1.00     1.00     0.00     1.00
PRECIP          1     10.00     1.00     1.00     1.00

Residuals, Leverage, and Influence of Observations
Leverage Limit                        0.154
Influence Limit                       0.103
         Observation  Residual  Leverage  Influence
                  1    0.139     0.265     0.042
                  2    0.290     0.133     0.091
                  3    0.112     0.199     0.021
                  4   -0.592     0.037     0.080
                  5   -0.301     0.055     0.033
                  6   -0.411     0.059     0.064
                  7   -0.293     0.027     0.019
                  8    0.271     0.112     0.068
                  9   -0.012     0.069     0.000
                 10    0.111     0.191     0.020
                 11    0.378     0.161     0.132
                 12   -0.359     0.022     0.031
                 13    0.189     0.024     0.009
                 14   -0.185     0.027     0.006
                 15   -0.176     0.087     0.019
                 16    0.091     0.138     0.009
                 17   -0.233     0.058     0.023
                 18   -0.373     0.030     0.037
                 19    0.198     0.034     0.009
                 20    0.245     0.011     0.009
                 21    0.422     0.011     0.030
                 22   -0.101     0.013     0.002
                 23    0.121     0.036     0.003
                 24   -0.011     0.138     0.000
                 25   -0.105     0.051     0.005
                 26    0.509     0.061     0.121
                 27    0.161     0.076     0.015
                 28    0.101     0.040     0.003
                 29    0.227     0.065     0.027
                 30    0.423     0.085     0.091
                 31    0.148     0.072     0.010
                 32   -0.456     0.087     0.102
                 33   -0.210     0.080     0.024
                 34   -0.264     0.118     0.057
                 35    0.100     0.073     0.005
                 36    0.061     0.050     0.001
                 37    0.563     0.026     0.098
                 38    0.236     0.054     0.023
                 39    0.221     0.125     0.045
```

Figure 17.   Example of output file ConventionalGLS.txt.

The observations are the observed flow characteristic (dependent variable) at each streamflow-gaging station and are numbered in ascending order, as the stations appear in the FlowChar.txt file (and other files).

RegionofInfluenceOLS.txt, RegionofInfluenceWLS.txt, and RegionofInfluenceGLS.txt

These three files show the same outputs but are produced using OLS, WLS, or GLS. The example shown in figure 18 is for RoI using OLS regression (RegionofInfluenceOLS.txt). This file provides information on the results of the RoI regression, and can be used to gauge whether an RoI regression may provide suitable results. However, WREG does not include a function that would allow users to perform a RoI regression at an ungaged location. The results shown by WREG are solely for demonstration or evaluation purposes.

WREG forms an individual RoI regression for each streamflow-gaging station included in the input dataset. The first several lines of the file show overall performance metrics for the RoI regression, when each of these individual

regressions are considered. Next, the coefficients of the RoI regression built for each site in the dataset are shown. If 200 sites are included in the dataset, 200 results will be shown— one for each regression that was built. These individual regressions will be numbered 1 to 200, and reflect the numerically ascending order of the Station IDs. That is, the Station ID with the smallest number would be associated with regression 1 or observation 1 in this output file and the Station ID with the largest number would be associated with the regression 200 or observation 200 in this output file.

The output file next displays the transformations that were used in the regressions. WREG does not allow different transformations to be used for different RoI regressions, and a single transformation applies to all individual RoI regressions formed by WREG. Next, the PRESS-like MSE of the RoI residuals are shown for the regression formed for each site in the dataset. Finally, the output file displays information on leverage and influence. For each regression, the output shows the observations that were used to form the regression as well as the leverage calculated for that observation. These outputs are paired. For example, in output shown in figure 18, the first regression used observations 2, 8, 3, 9, 14, 13, 12, 7, 10, and 5. Following each observation number,

```
Region-of-Influence Regression Models for Q1%

Performance Metrics
Root Mean Square Error              73.45
Pseudo R2                           NA
Standard Model Error                NA
*Note: Root Mean Square Error is in percent.

Coefficients of Model
Regression    Constant      Coefficients
                            DRNAREA      PRECIP
           1      4.61         0.54       -0.01
           2      3.84         0.54       -0.00
           3      4.10         0.29       -0.00
           4     -2.65         0.60        0.02
           5      2.97         0.30       -0.00
           6     -1.28         0.70        0.01
           .
           .
           .

Transformations            C1       C2       C3       C4
Q1%           log10       1.00     1.00     0.00     1.00
DRNAREA       log10       1.00     1.00     0.00     1.00
PRECIP            1      10.00     1.00     1.00     1.00

PRESS_RoI Residuals
Observation                Residual
           1                 0.060
           2                 0.028
           3                 0.141
           4                 0.262
           5                 0.047
           6                 0.132
           .
           .
           .

Leverage and Influence
Leverage Limit                  0.400
Influence Limit                   NA
Regression        Observations Included in RoI
 Model                 and Leverage Value
         1    2  0.400    8  0.293    3  0.461    9  0.184   14 -0.038   13 -0.046   12 -0.075    7 -0.163   10  0.149    5 -0.165
         2    1  0.315    8  0.169    3  0.356   13  0.050   14  0.020   12  0.035    9  0.054    7 -0.012    5  0.035   18 -0.020
         3    2  0.385    1  0.445    8  0.112   11  0.518   13 -0.081   12 -0.113    5 -0.046   14 -0.202   17 -0.041   35  0.024
         4   23  0.096   22  0.106    7  0.083   14  0.099   20  0.120    6  0.133   12  0.080   18  0.070   19  0.133   13  0.079
         5   17  0.249   18  0.068   35  0.327   12  0.141    7  0.031   13  0.172   36  0.033   25 -0.080   32  0.023   14  0.035
         6   15  0.159   19  0.098    4  0.073   20  0.077   22  0.062   14  0.086   23  0.051   16  0.174    9  0.172    7  0.047
         .
         .
         .
```

Figure 18. Example of output file RegionofInfluenceOLS.txt.

the leverage value for that observation is shown. For observation 2, the leverage was 0.400.

RegressionModel.txt

The RegressionModel.txt file gives the regression equation calculated by WREG. Transformations, variables, and parameters are all shown. In the example shown for conventional OLS (figs. 13 and 16), the corresponding regression equation contained in RegressionModel.txt is shown in figure 19. This equation can be interpreted as

$$\log(Q1\%) = 0.97 + 0.51 \log(DRNAREA) + 0.01(10Precip + 1). \quad (48)$$

```
log 10{[1.00*(Q1%)^(1.00)+0.00]^(1.00)}  1.15
+0.52*log 10{[1.00*(DRNAREA)^(1.00)+0.00]^(1.00)}
+0.00*[10.00*(PRECIP)^(1.00)+1.00]^(1.00)
```

Figure 19. Example of regression model equation shown by RegressionModel.txt.

This output file is produced only for conventional regressions.

InvXLX.txt

This file contains the covariance matrix of the regression parameters (equation 46). Its size depends on the number of dependent variables used in the regression. An example is shown in figure 20. This matrix can be used to estimate the variance of individual regression estimates, which varies depending on the values of basin characteristics. This output file is created only for conventional regressions.

```
 6.0173594e-001   1.4652262e-003  -1.8692077e-003
 1.4652262e-003   2.2218603e-003  -1.4027835e-005
-1.8692077e-003  -1.4027835e-005   5.8799320e-006
```

Figure 20. Example of InvXLX.txt output file.

SSres.txt and SStot.txt

The SSres.txt file reports the residual sum of squares (equation 36) and the SStot.txt file reports the total sum of squares (equation 37). These output files are created only for conventional regressions. Examples are shown in figure 21.

```
3.0515284e+000
```
```
1.3821829e+001
```

A   *B*

Figure 21. Example of *A*, SSres.txt and *B*, SStot.txt output files.

EventLog.txt

The event log contains a record of the WREG session that can be useful for debugging. If the program is operating normally, it does not need to be considered.

Other Program Notes

Screen Capture.—As noted on many windows, the contents of that window can be captured by pressing Alt and the Print Screen key. This will copy an image of the window to the clipboard. It needs to be pasted into a document (for example, a WordPad document) in order to save it.

Acknowledgments

The authors would like to thank Jery R. Stedinger (Cornell University), Andrea M. Gruber-Veilleux (Cornell University), and Charles Parrett (USGS California Water Science Center) for their helpful comments on this report.

Future updates will be available at *http://water.usgs.gov/software/WREG/*.

References Cited

Acreman, M.C., and Wiltshire, S.E., 1987, Identification of regions for regional flood frequency analysis (abstract): EOS, v. 68, no. 44.

Aitchison, J., and Brown, J.A.C., 1957, The lognormal distribution: Cambridge University Press, 176 p.

Burn, D.H., 1990, Evaluation of regional flood frequency analysis with a region of influence approach: Water Resources Research, v. 26, no. 10, p. 2257–2265.

Cook, R.D., 1977, Detection of influential observation in linear regression: Technometrics, v. 19, p. 15–18.

Eng, Ken, Tasker, G.D., and Milly, P.C.D., 2005, An analysis of region-of-influence methods for flood regionalization in the Gulf-Atlantic Rolling Plains: Journal of American Water Resources Association, v. 41, no. 1, p. 135–143.

Eng, Ken, Milly, P.C.D., and Tasker, G.D., 2007a, Flood regionalization: a hybrid geographic and predictor-variable region-of-influence regression method: Journal of Hydrologic Engineering, ASCE, v. 12, no. 6, p. 585–591.

Eng, Ken, Stedinger, J.R., and Gruber, A.M., 2007b, Regionalization of streamflow characteristics for the Gulf-Atlantic Rolling Plains using leverage-guided region-of-influence regression, Paper 40927–3050, *in* Kabbes, ed., Restoring our natural habitat—Proceedings of the World Environmental and Water Resources Congress, May 15–18, 2007, Tampa, Florida: American Society of Civil Engineers.

Funkhouser, J.E., Eng, Ken, and Moix, M.W., 2008, Low-flow characteristics for selected streams and regionalization of low-flow characteristics in Arkansas: U.S. Geological Survey Scientific Investigations Report 2008–5065, 161 p., available only online at *http://pubs.usgs.gov/sir/2008/5065/.* (Accessed August 19, 2009.)

Giese, G.L., and Mason, R.R., Jr., 1993, Low-flow characteristics of streams in North Carolina: U.S. Geological Survey Water Supply Paper 2403, 29 p., 2 plates. (Also available online at *http://pubs.er.usgs.gov/usgspubs/wsp/wsp2403.*)

Griffis, V.W., and Stedinger, J.R., 2007, The use of GLS regression in regional hydrologic analyses: Journal of Hydrology, v. 344, p. 82–95.

Griffis, V.W., and Stedinger, J.R., 2009, Log-Pearson type 3 distribution and its application in flood frequency analysis. III: sample skew and weighted skew estimators: Journal of Hydrologic Engineering, v. 14, no. 2, p. 121–130.

Hardison, C.H., 1971, Prediction error of regression estimates of streamflow characteristics at ungaged sites: U.S. Geological Survey Professional Paper 750–C, p. c228–c236. (Also available at *http://pubs.er.usgs.gov/usgspubs/pp/pp750C.*) (Accessed August 19, 2009.)

Hirsch, R.M., 1982, A comparison of four streamflow record extension techniques: Water Resources Research, v. 18, no. 4, p. 1081–1088.

Interagency Advisory Committee on Water Data, 1982, Guidelines for determining flood-flow frequency, Bulletin 17B of the Hydrology Subcommittee, Office of Water Data Coordination: U.S. Geological Survey, Reston, Va., 183 p. (Also available at *http://water.usgs.gov/osw/bulletin17b/dl_flow.pdf.*) (Accessed August 19, 2009.)

Kenney, T.A., Wilkowske, C.D., and Wright, S.J., 2007, Methods for estimating magnitude and frequency of peak flows for natural streams in Utah: U.S. Geological Survey Scientific Investigations Report 2007–5158, 28 p. (Version 4.0 was released March 10, 2008, and is available only online at *http://pubs.usgs.gov/sir/2007/5158/.*) (Accessed August 19, 2009.)

Kite, G.W., 1975, Confidence limits for design events: Water Resources Research, v. 11, no. 1, p. 48–53.

Kite, G.W., 1976, Reply to comment on Confidence limits for design events: Water Resources Research, v. 12, no. 4, p. 826.

Martins, E.S., and Stedinger, J.R., 2002, Cross correlations among estimators of shape: Water Resources Research, v. 38, no. 11, p. 34-1–34-7, doi:10.1029/2002WR001589.

Merz, R., and Blöschl, G., 2005, Flood frequency regionalisation—spatial proximity vs. catchment attributes: Journal of Hydrology, v. 302, p. 283–306.

Montgomery, D.C., Peck, E.A., and Vining, G.G., 2001, Introduction to linear regression analysis (3d ed.): New York, John Wiley and Sons, 641 p.

Neely, B.L., Jr., 1986, Magnitude and frequency of floods in Arkansas: U.S. Geological Survey Water Resources Investigations Report 86–4335, 51 p., available only online at *http://pubs.er.usgs.gov/usgspubs/wri/wri864335/.* (Accessed August 19, 2009.)

Ries, K.G., III, and Fries, 2000, Methods for estimating low-flow statistics for Massachusetts streams: U.S. Geological Survey Water-Resources Investigations Report 2000-4135, 81p. (Also available at *http://pubs.usgs.gov/wri/wri004135/.*) (Accessed August 24, 2009.)

Ries, K.G., III, comp., 2007, The National Streamflow Statistics Program: A computer program for estimating streamflow statistics for ungaged sites: U.S. Geological Survey Techniques and Methods book 4, chap. A6, 45 p. (Also available at *http://pubs.usgs.gov/tm/2006/tm4a6/.*) (Accessed August 19, 2009.)

Stedinger, J.R., and Tasker, G.D., 1985, Regional hydrologic analysis, 1, ordinary, weighted, and generalized least squares compared: Water Resources Research, v. 21, no. 9, p. 1421–1432.

Stedinger, J.R., and Tasker, G.D., 1986, Regional hydrologic analysis, 2, model-error estimators, estimation of sigma and log-Pearson Type 3 distributions: Water Resources Research, v. 22, no. 10, p. 1487–1499.

Tasker, G.D., 1980, Hydrologic regression with weighted least squares: Water Resources Research, v. 16, no. 6, p. 1107–1113.

Tasker, G.D., and Stedinger, J.R., 1986, Regional skew with weighted LS regression: Journal of Water Resources Planning and Management, ASCE, v. 112, no. 2, p. 225–236.

Tasker, G.D., and Stedinger, J.R., 1989, An operational GLS model for hydrologic regression: Journal of Hydrology, v. 111, p. 361–375.

Tasker, G.D., Hodge, S.A., and Barks, C.S., 1996, Region of influence regression for estimating the 50-year flood at ungaged sites: Water Resources Bulletin, v. 32, no. 1, p. 163–170.

Thomas, D.M., and Benson, M.A., 1970, Generalization of streamflow characteristics from drainage-basin characteristics: U.S. Geological Survey Water-Supply Paper 1975, 55 p. (Also available online at *http://pubs.er.usgs.gov/usgspubs/wsp/wsp1975.*) (Accessed August 19, 2009.)

Wandle, S.W., 1977, Estimating the magnitude and frequency of floods on natural-flow streams in Massachusetts: U.S. Geological Survey Water-Resources Investigations Report 77–39, 26 p., 1 plate. (Also available online at *http://pubs.er.usgs.gov/usgspubs/wri/wri7739.*)

www.ingramcontent.com/pod-product-compliance
Lightning Source LLC
Chambersburg PA
CBHW081420170526
45166CB00010B/3416